Use of Fermentation Technologies: A Biological Perspective

USE OF FERMENTATION TECHNOLOGIES: A BIOLOGICAL PERSPECTIVE

Edited by:

Dr. Sujata Mustapure

ARCLER PRESS

www.arclerpress.com

Use of Fermentation Technologies: A Biological Perspective

Dr. Sujata Mustapure

Arcler Press
224 Shoreacres Road
Burlington, ON L7L 2H2
Canada
www.arclerpress.com
Email: orders@arclereducation.com

ISBN: 978-1-77469-865-5 (Hardcover)

ABOUT THE EDITOR

Dr. Sujata Nagnath Mustapure (1988) is presently serving as teaching associate, Department of Electrical & Other Energy sources, College of agricultural engineering & Technology, Vasantrao Naik Marathwada Krishi Vidhyapeet, Parbhani, Maharashtra. She obtained her B.Tech (agril. Engg.) in 2010 from Vasantrao Naik Marathwada Krishi Vidhyapeet, Parbhani. Gold medal in M.Tech (Agril. Engg.), 2014 from Dr. Panjabrao Deshmukh Krishi Vidyapeeth, Akola and Ph.D in Renewable energy engineering in 2019 from Maharana Pratap University of Agriculture and Technology, Udaipur. She has also worked as assistant professor in College of Agricultural, Naigoan, Vasantrao Naik Marathwada Krishi Vidhyapeet, Parbhani.

TABLE OF CONTENTS

LIST OF FIGURES

PREFACE

The main agents in the fermentation process that produces enzymes, organic acids, alcohols, polymers, and other products with added value are microbes. It is a microbial biotechnology procedure. Today, targeted genetic engineering approaches applied to establish industrial microbial strains have profited from increased production of economically significant fermentation products. Due to their catalytic activity, reproducibility, etc., enzymes, which are by-products of living organisms, have been utilized in the industry for a long time. A wide range of products, including food and detergents use enzymes. The majority of enzymes are marketed and used to improve the quality of goods such as food, detergents, leather, paper, cosmetics, and medications or drugs. A highly interesting group of culinary products is fermented food. In addition to being easily made using low-cost ingredients and basic methods, it also significantly contributes to the human diet, particularly in rural communities in many parts of the world. Fermented foods are prepared by every ethnic group in the world using recipes that have been passed down for generations.

These food items contribute significantly to local economies, cultural identities, and culinary preferences. Some of the more popular fermented food products are now manufactured commercially, thanks to advancements in biological and microbiological sciences, while others are still made using conventional techniques at home or in small-scale businesses. The basic features of a food article are altered throughout fermentation to create a product that is different but still quite popular with people. Consumer preference for fermented food differs both within and between cultures. Fermentation was once used to increase the shelf life of foods as well as their texture and sensory qualities. The early civilizations were also aided in their efforts to unite and socialize by a large number of fermented foods and drinks. Numerous civilizations have recognized the nutritional and health benefits of including fermented foods and beverages in a regular diet. From a health and dietary perspective, fermented foods have also attracted attention in recent days. To ensure that the process is streamlined and the final product is of high quality, scaling up and optimizing food goods is a necessity for the commercialization of food products. Process optimization for fermented foods and beverages is difficult because of the dynamic nature of fermentation and the variety of fermentation microorganisms.

For researchers, teachers, and students, this book will be a valuable reference source on fermentation and food microbiology.

Introduction

The products made from fermented cereals and other crops offer calories, vitamins, and fiber. They contribute lactic acid bacteria that may function similarly to probiotics. These foods may protect for newborns, children, and other vulnerable populations, particularly in areas where deadly diarrheal infections are prevalent. These products are typically poor in lysine content, an important amino acid, and minerals that could not be available depending on the substrate. Because of this, even though these foods have many good qualities, it is important to carefully consider whether they should be used as the only source of nutrition for children. The sensory quality, nutritional value, and possibly safety of fermented cereal products vary often despite their traditional, domestic production methods. As a result, there is interest in creating starting cultures for fermented cereals and other crops to meet customer expectations and produce more consistent quality. In addition to the potential probiotic effects of each strain, some strains of lactic acid bacteria and related microorganisms like Bifidobacteria require some enzymes to produce fermented or pre-treated soymilk and others with improved properties i.e., low flatulence, good digestibility, enrichment in isoflavones, etc.

Galactosidase helps to make soymilk and others with low levels of α-GOS, which causes digestive problems in people. Soy products and other milk products with better digestibility and bioactive peptides that prevent cancer, promote bone health, lower cholesterol, and have anti-aging and antioxidant properties are made possible by proteases and peptidases.

The bioconversion of isoflavones into aglycones, which may be ingested by people and have positive effects on bone mineralization, the relief of menopausal symptoms, the hypocholesterolemic effect, and antiaging/antioxidant effects, is facilitated by β-glucosidases. Fermented goods with health advantages are gaining attention, and those foods whose health benefits are sufficiently substantiated by scientific studies have the potential to be a vital part of a healthy lifestyle.

The fundamentals of fermentation technology, as well as its function in safeguarding people from diseases and protocols related to the fermentation process, are covered in this book. Students who want to understand the fundamentals, as well as more complex ideas related to fermented foods and their manufacturing, will benefit from this content.

CHAPTER

1 Introduction to Fermentation Technology

Contents

1.1 INTRODUCTION

Fermentation technology (Figure 1.1) refers to a field that uses microbial enzymes to produce molecules with uses in the chemical, food, pharmaceutical, energy, and pharmaccutical industries (Formenti et al., 2014). People have used it to preserve food and enhance its organoleptic qualities (Pua et al., 2022). It is a technology from the prehistoric era that was used to preserve food and produce bread, beer, vinegar, yogurt, cheese, and wine. The production of enzymes, metabolites, biomass, recombinant technologies, and biotransformation products on an industrial scale are all examples of this biological process, which involves a variety of microorganisms like yeast, bacteria, and fungi. The primary by-products of fermentation are organic acid and alcohol. Secondary metabolites i.e., antibiotics, enzymes, and growth hormones are released during this process (Robinson et al., 2001). They become biologically active and so they are called bioactive molecules (Stepek & Bode, 2018). Secondary metabolites are the building blocks of many bioactive substances, including phenolic compounds, growth hormones, food colors, antibiotics, mycotoxins, and alkaloids. Flavonoids, tannins, and phenolic acids are the components of phenolic compounds. Some important classes of flavonoids include flavanones, flavonols, flavones, anthocyanidins, and isoflavones.

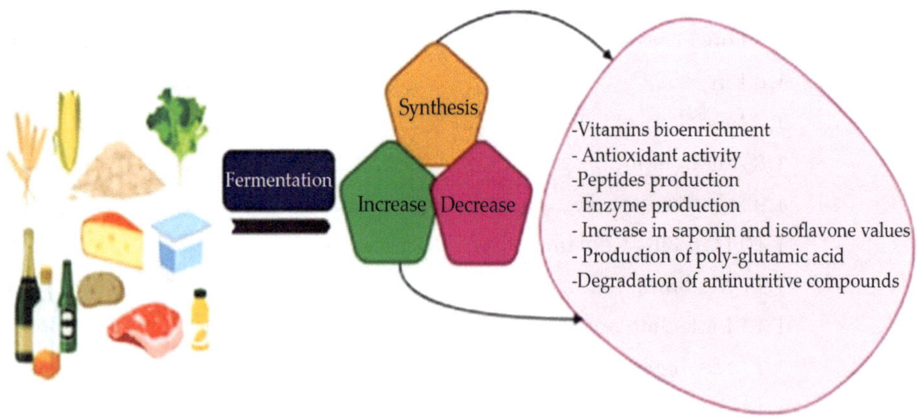

Figure 1.1. Fermentation technology.

(**Source:** *Capozzi V, Fragasso M, Bimbo F. Microbial Resources, Fermentation and Reduction of Negative Externalities in Food Systems: Patterns toward Sustainability and Resilience. Fermentation. 2021; 7(2):54. https://doi.org/10.3390/fermentation7020054, Creative Commons License)*

Typically, a bioreactor is needed in this process. Submerged fermentation (SmF) involves the mixing of substrates in both single and mixed cultures where there is an adequate supply of oxygen(Hansen et al., 2015). Solid-state or substrate fermentation (SSF) is defined as fermentation taking place in a natural environment with minimal moisture content and a solid supporting medium (Singhaniaet al., 2009). Substrates like garbage that are rich in nutrients can be recycled in this procedure. In SSF, solid substrates including paper pulp, bran, and bagasse are employed.

1.2 FERMENTATION

The fermentation of the substrate requires a lot of time because the process is slow. Therefore, the release of nutrients is controlled. As a result, it is the fermentation method used by fungi and microorganisms since it requires minimal moisture content. Microorganisms in SmF needed a controlled environment to efficiently produce high-quality end products and achieve maximum productivity and yield. Industrial bioreactors use batch, fed-batch, or continuous modes for the production of a variety of microorganisms. Preservatives or acidifiers like lactic acids, citric acid, and vinegar are employed in the food sector for the production of alcoholic drinks like whisky, beer, brandy, rum, and wine, and submerged batch cultivation is used for flavor enhancers like monosodium glutamate or sweeteners like aspartate (Akabandaet al., 2014). A closed system called batch culture operates in an aseptic environment. The bioreactor in which the amount of the culture broth is constant is used to combine the inoculums, nutrients, and medium. Because the fermentation product might vary, it is crucial to choose an appropriate substrate. Every substrate is optimized and this is mostly because microorganisms respond differently to each substrate.

Every substrate has a varied rate of nutrient consumption, which affects production. Rice straw, vegetable waste, wheat bran, fruit bagasse, and paper pulp are a few of the substrates that are frequently used in SSF (Dhillon et al., 2011). To extract bioactive chemicals, liquid media, molasses, wastewater, vegetable juices, and soluble sugar are often employed as substrates in SmF. Several bioactive chemicals are extracted by fermentation, including enzymes, antioxidants, antibiotics, biosurfactants, and pigments. The most crucial method for producing various enzymes is enzyme culture. Both fungus and bacterial microorganisms are necessary for fermentation to take place on the substrates to produce the enzymes. Submerged and SSF can work together to produce enzymes. Due to the high-water need, the

SmF technique is frequently used in the manufacture of bacterial enzymes. SSF technique is used in fungus because less water is needed. Seventy-five percent of all industries use SmF to produce enzymes worldwide. It is better to prefer SmF over SSF because it supports genetically modified organisms (GMOs) to a greater extent than SSF does (Holker et al., 2004). Various metabolic parameters of cultivation are used to carry out nutrient inflow and waste efflux. A slight deviation from the specified parameters will have an impact on the product. The bacterial enzymes cellulase, amylase, xylanase, and L-asparaginase are only a few examples (Kango et al., 2019). In the past, SmF was one of the finest methods for generating enzymes from bacteria. Recent research has revealed that SSF is more effective than SmF at producing bacterial enzymes. The SmF system uses a collection of several intermediary metabolites to reduce enzyme activity (Hartmanis & Gatenbeck, 1984).

Figure 1.2. Aspergillus.

(**Source:** *https://upload.wikimedia.org/wikipedia/commons/6/68/Aspergillus_parasiticus_UAMH3108.jpg, Creative Commons License)*

1.3 FUNGI

Aspergillus fungus (Figure 1.2) has long served as a popular example of a model microorganism for the fungus enzyme manufacturing process. One of the main sources of fungal enzymes is Aspergillus. The primary distinction between SSF and SmF is how productive the fungus is. Thermoascusauranticus phytase is isolated using SmF (Tanruean et al., 2021). Antibiotics are significant microorganism extracts produced utilizing fermentation technology. The first antibiotic made through fermentation is penicillin, which comes from Penicillium notatum (Figure 1.3).

Figure 1.3. Penicillium notatum.

*(**Source:** https://maas-website-media.s3.amazonaws.com/maas.museum/uploads/sites/4/2018/10/penicillium-notatum-e1539580800317-1024x729.jpg, Creative Commons License)*

Penicillins, Cephalosporins, Monobactams, Aminocillins, and Carbapencins collectively are referred to as beta-lactam antibiotics. This technique is used to produce several different antibiotics, including Tetracycline, Streptomycin, Cyclosporin, Cephalosporin, and Surfactin. Cephamycin C is produced by Streptomyces clavuligerus, Nocardialactamdurans, and Streptomyces cattleya from sunflower cake and cotton-deoiled cake as substrates (Bignell et al., 2005). In SSF, actinomycetes and fungi in mixed cultures produced penicillin.

Figure 1.4. Bioreactor.

(Source: https://www.engr.colostate.edu/CBE101/_images/bioreactor_industrial.png, Creative Commons License)

Nowadays, SSF is used more frequently than SmF due to the development of suitable substrates. Some findings indicate that particular microbiological stains are more appropriate for SmF and others are more appropriate for SSF. As a result, the development of this technology is based on the microorganisms utilized. Recent research has shown that certain antibiotics produced by SSF are more consistent than those produced by SmF. SSF implementation is constrained by the properties of the substrate material. This characteristic makes it necessary to evaluate the production potential of various substrates before fermentative process optimization.

1.4 BIOREACTOR

Typically, the bioreactors (Figure 1.4) are filled with sterilized media at the start of batch production and are inoculated with a known quantity of viable cells (Gallazzi et al., 2015). Alcoholic beverages, organic acids used as preservatives or acidifiers, and sweeteners or amino acids used as flavoring agents are products produced by submerged batch cultivation (Bakratsas et al., 2021). In fed batch cultivation, one or more nutrients are added aseptically, the system is semi-open, and the culture is supplemented into the bioreactor step-by-step while the amount of the liquid culture in the bioreactor increases during this time. The key benefits of fed-batch cultures over batch cultures include increased productivity, improved yield from the controlled addition of nutrients, the ability to reach higher cell densities, and product synthesis. Most research on cell immobilization occurs in the food and gas-liquid mass. The majority of microbial enzymes are produced using this method. The manufacture of important industrial enzymes such as cellulases, lipases, proteases, glucoamylases, amylases, ligninases, xylanases, pectinases, and peroxidases has reportedly been the subject of investigation. This technique has resulted in the production of the thermophilic Bacillus licheniformis (Figure 1.5) enzyme xylanase (Archana & Satyanarayana, 1997).

This approach results in more thermo-stable enzymes than the SmF process. SSF produces several acids, including lactic acid, fumaric acid, kojic acid, and gallic acid. Some agro-industrial wastes that are extremely resourceful substrates for the generation of citric acid in SSF include wheat bran, de-oiled rice bran, sugarcane, coffee husk, kiwi fruit peels, pineapple wastes, grape pomace, and apple. Pineapple waste was used as a substrate for the synthesis of citric acid (Imandi et al., 2008). Rhizopus oryzae is utilized to produce lactic acid, and sugarcane bagasse that has been impregnated with glucose and $CaCO_3$ is employed in the process.

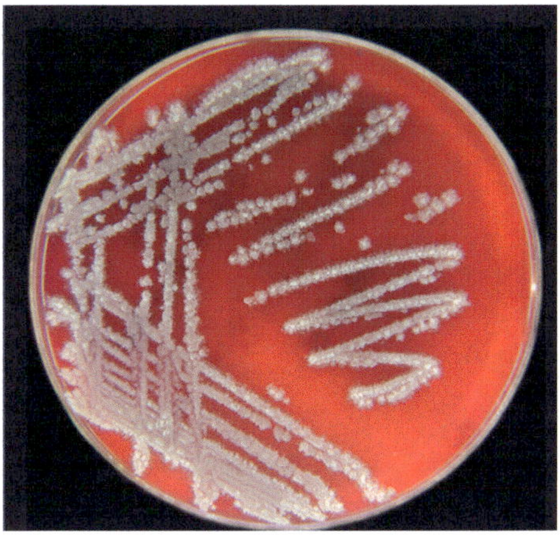

Figure 1.5. Bacillus licheniformis.

*(***Source:** *https://upload.wikimedia.org/wikipedia/commons/6/61/Bacillus_licheniformis.jpg, Creative Commons License)*

Gibberellic acid is a secondary metabolite produced by fungi during their stationary phase. In the SSF system, gibberellic acid synthesis rises. Gibberella fujikuroi was employed in the manufacture of gibberellic acid with wheat bran as the substrate, and the amount of gibberellic acid collected was 1.626 times more in SSF than SmF(Kumar &Lonsane, 1987). Common antibiotics made from SSF include Cephamycin C, Cyclosporin A, Penicillin, Neomycin, Iturin, and Cephalosporins. Penicillium chrysogenum, which produces penicillin, is grown on wheat bran and sugarcane bagasse under high moisture conditions. Cephamycin C is produced by Nocardia lactamdurans, Streptomyces clavuligerus, and Streptomyces cattleya. Penicillin, an antibiotic, is made in SSF using mixed cultures of Actinomycetes and fungi. Ethanol is currently the most popular biofuel. Even though it is fairly simple to create ethanol using SmF, it is preferred due to its low water demand, end product protection that is inhibited, less liquid water disposal, reduced pollution issue, and widespread availability. For the controlled fermentation of SSF of apple pomace mixed with ammonium sulfate, Saccharomyces cerevisiae (Figure 1.6) is employed. Substrates that are frequently used include sweet potatoes, rice starch, wheat flour, potato starch, and sweet sorghum.

Figure 1.6. Saccharomyces cerevisiae.

(**Source:** *https://upload.wikimedia.org/wikipedia/commons/9/95/Saccharomyces_cerevisiae_SEM.jpg, Creative Commons License)*

1.5 FUNGAL AGENTS

Fungal agents have a larger potential to operate as biocontrol agents due to their various modes of action. Liagenidiumgiganteum is used as a fungus to control insects (Kerwin, 2007). The water-soluble enzymes produced on SSF by various species of Rhizopus and Klebsiella, a well-known producer of vitamin B12, include nicotinic acid, thiamine, riboflavin, and vitamin B6. Microbial enzymes are widely utilized in the food industry. For the formation of new products, numerous novel industrial and analytical applications have been developed recently. The food and beverage sectors require the improvement of fermentation technology. The development of contemporary instruments and sensors for process monitoring is highly acceptable in SSF. This method is inappropriate for large-scale applications due to its main disadvantage i.e., high cost. Microorganisms use the metabolic process of fermentation to produce a variety of products. To create specific goods, different species require various environmental factors. For effective production, several variables like biomass resource, type of microbe, growth rate, agitation speed, substrate composition, reaction duration, and pH of the culture medium must be optimized. This process is known as simultaneous saccharification and fermentation (SSF). The issue is thus resolved through optimization using modeling and kinetics. Kinetics is the study of how measurements are interpreted and the variables that affect the fermentation process. Three basic approaches i.e., phenomenological, thermodynamic, and kinetic can be used to explain this analysis. Mathematical representations are the most effective in interpreting the system's dynamics and modelling (MacLeod & Nersessian, 2015). An optimized model of the fermentation system for the production of a specific product can be made by manipulating and converting mathematical expressions observingthe requirements of the system. Modeling the kinetics of the fermentation process is one of the most crucial aspects of studying the fermentation process since it aids in process control and research. It efficiently lowers production costs and boosts the production of more products of higher quality.

1.6 KINETICS

The performance of the bioreactor is taken into account while modeling the fermentation process, in addition to the kinetics of the cell system. Microbial kinetics and bioreactor performance are thus the two components of modelling (Demirkaya et al., 2022). As the fermentation process now depends on several variables, including temperature, aeration, substrate, biomass, etc., the production of products is affected. Low productivity results from the lack of ideal sensors for quantifying product production, substrate, and biomass, and manually tuning the system is difficult. As a result, to boost productivity, other parameters that affect the fermentation process must be managed, which increases the demand for labor and the cost of production. Therefore, automation of the fermentation systems is required to reduce costs. Models depict the interactions between the system and the variables that have an impact on it. Any piece of unit-operating equipment, including bioreactors, individual cells, microbial cultures, immobilized cells, HPLC, etc., can be considered a system.

Different variables of interest, such as time, temperature, reaction rate, etc., have an impact on a system. System modeling helps to improve the conditions that will lead to greater system performance. In the case of fermentation, numerous variables can affect the system and the environment, including feed rate, pH, the rate and mode of agitation, inoculum quality, temperature, prices of production systems, etc. Modeling might be carried out utilizing experimental techniques or non-mathematical expressions. The best method is mathematical modeling because it predicts the system's outcome without carrying out the real experiment. Contrarily, non-mathematical experimental approaches are laborious since they require a lot of time, and are unpredictable. Since mathematical modeling can foresee outcomes beforehand, it reduces the cost of system modeling. The process of modeling a system is iterative and involves numerous variables that must be kept under close watch. To enhance the model and create an optimum model for the particular system, more studies are to be conducted.

1.7 MODELING

Since modeling is an iterative process with multiple optimization steps, we can begin by considering any basic components, such as a set of data from a batch culture. Another example of modeling the culture system is changing the culture ingredients and tracking the rate of cell growth and the generation of extracellular components over time. Modeling a system involves altering a parameter that is involved in it. Mathematical modeling is the best modeling approach. Now, mathematical formulas create a dynamic model of the system when combined with the power of automation. Automatization is accomplished with the aid of computer systems (Borah & Bhuyan, 2005). There is software available now that can quickly analyze and understand data. Computer-based fermentation system modeling has increased output (Liu et al., 2001). Huge amounts of quantitative and qualitative data are created in fermentation processing and biotechnology engineering as a result of technological advancements in measurement, instrumentation, information technology, molecular biology, and high-throughput approaches. These data are processed, and various tools are used to know the relationships and connectivity between them. A model is created once the relation and connectivity have been identified. The best approach is to build a hypothesis before building a model because modeling is a cyclical process. Thus, modeling offers information that may be used to forecast how the fermentation system would behave. It forecasts the results before an experiment is conducted. In this approach, depending on our requirements for the formation of the product, we can select the ideal model or build a new model using the existing model. It lowers the cost of labor or manpower, and system automation ensures error-free analysis and minimal loss.

It also serves as a common medium for scientists and engineers from various backgrounds by employing model-based terminologies. Without having to repeat trials, it makes it easier to accurately predict and choose the next experiment. The model automatically measures, monitors, and occasionally highlights variables that are thought to be of lower value but are crucial to the fermentation system. Understanding the elements of modeling is necessary for building a model. Knowing the modeling parameters aids in system prediction. A model is considered accurate when it can accurately predict the results of a series of independent trials that include various experimental variables, such as those related to the fermentation system. Experimental error and physical constraints should also be considered while building a model (Wang et al., 2001). Leaving out data with a high degree of error is one type of experimental error. As a result, the model ought to include replications of experiments, sampling, and analysis. Physical constraints are the limit boundaries of the range of values of the system variables and parameters, which must be taken into consideration. They also include technological, biological, chemical, and physical constraints.

One of the most crucial parts of modeling is the control zone. The area in the system known as the control region is where all of the system variables such as concentration, pH, temperature, and pressure are maintained uniformly. The concentration in the control zone does not have to remain constant over time. Concentration may change over time or stay the same, any change in the control region stays consistent. This indicates that the concentration of the system components, for instance, remains constant over time in the control region. Since the majority of systems are heterogeneous, modelers typically view the control zone as an idealized version of the system. Depending on the degree of homogeneity, more than one control region is taken into account while dealing with a heterogeneous system. The bulk liquid, which serves as the control region in bioreactors where concentrations are uniform throughout the system, makes the control region. However, if the impeller in a bioreactor divides the compound concentration into two or more halves while maintaining uniform compound concentration in each half, the system will have more than two control regions. The control region is the bulk liquid. With the control areas, there may be an interchange of matter, energy, or momentum. The control region's volume could change or stay the same. Both finite and infinitesimal control zones are possible. A few of the boundaries in control regions can be categorized as phase boundaries across which no exchange occurs, phase boundaries across which an exchange of mass and energy occurs, and geometrically defined boundaries within a single phase where the exchange occurs by bulk flow or molecular diffusion. A critical stage in the modeling process for the model's success is selecting the control volume. Although it appears to be simple, some numerous variables and aspects must be taken into account, making the procedure complex. Thus, it can be said that to build a model, one should first assume the system to be developed, then take into account the mode of operation or activity, which helps determine whether the system will be stable or unsteady, that is, if the system properties should vary or not.

The system heterogeneity also determines whether the control zone will be infinitesimal or finite(Dedey et al., 2021).

1.8 AUTOMATION

Computer automation (Figure 1.7) has permeated every industry sector, displacing human labor power. It has produced more accurate and error-free processes. Automation has already flourished in sectors like the oil industry, metallurgy, chemical industry, etc., whereas fermentation industries took much longer to succeed. The causes of this include a lack of adequate sensors for products, substrates, and biomass; a lack of dependable process models for process control analysis; and higher investment costs for computers in the fermentation field compared to other industries because fermentation was previously a small-scale production. But nowadays, large-scale fermentation companies are expanding quickly, and improved sensors and fermentation models are available to speed up the fermentation process.

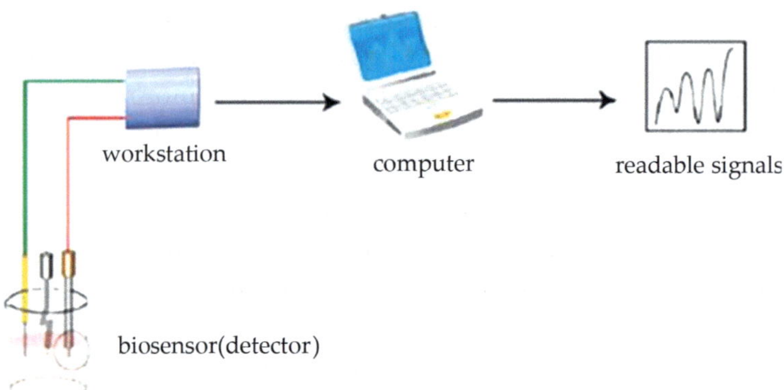

Figure 1.7. Computer automation.

*(**Source:** Shi, J., Feng, D., & Li, Y. (2017). Biosensors in Fermentation Applications. Fermentation Processes. doi: 10.5772/65077, Creative Commons License)*

A fermenter is a vessel with controlled temperature, aeration, and pH where bacteria are cultivated for the fermentation process. It has an input port and an output port. To control the input of the substrates concerning the output of the product, a sensor could be connected to the output port. In this manner, a fermenter may be automated, and the entire process could be monitored and sensed by a computer system. A computer is a digital machine that can carry out a task automatically by carrying out operations by a specified set of variables and programs that have been allocated to it. A computer should have software and applications that can interpret the data produced by a fermentation system and recreate it understandably. Therefore, a powerful computer with increased storage capacity and fast processing speed may be the most appropriate for automating the fermentation system.

An input and output port on a computer system is used to send and receive data and control signals between the computer and the fermenter. The parallel port and serial port are the two basic types of ports. While a serial port sends data bits one at a time, a parallel port sends data bits simultaneously. The process operator console and process computer connect over a serial port. The serial port helps control data transfer between the computer and the terminal. Similar to serial ports, where a single link is sufficient to send all data bits across vast distances, parallel ports require as many links as there are data bits to be transferred. Parallel lines are primarily employed to connect the computer to the processing system. They serve as input and output ports for sensors, control lines, and data transfer. Without utilizing a costly interface, a switch can be connected to a computer using a parallel port. However, a buffer is required only in case the system doesn't crash from being overloaded. To handle the high voltage or current, several control valves are also required. To convert to digital format when using an analog feed, an analogue-to-digital converter is required. When a computer is connected to the entire fermentation system, the question of when, how, and how much data should move through the input and output ports becomes relevant. The system needs to be connected to a computer and some floating sensors, such as a pH meter and a spectrometer, so the computer can analyze and compare the data produced and turn the system into a fully automated one. In non-computerized fermentation, an analog controller manages the process operations (De Buyser et al., 1985). As a result, this process system has an analogous regulatory mechanism that maintains the variables under control at a programmed level. As these control points are manually set, they must be manually adjusted numerous times if they are not consistent. Set point control is the term given to this manual setting process. Computers are also capable of performing controlling tasks. As a result, by comparing the data generated by sensors with the value stored in the computer's memory, it is possible to determine whether to take a combined or different course of action for the fermentation system. Direct digital control is the process of connecting the fermentation system with computers. With the expansion of knowledge in recent years, information about metabolic processes, product formations, cell development and death, changes in their morphology, etc., has been produced. They have an impact on the fermentation system as well. As a result, the biological system can be illustrated via mathematical models. As a result, every biological response is regarded as having its reaction system. Cell maintenance or endogenous respiration is regarded as fundamental cell function (Garcia-Ochoa et al., 2010). Another presumption is that the entire population of living cells in the microbial culture constitutes a single homogeneous biomass and that all microbial cells have the same physiology i.e., shape, size, etc. Understanding the kinetic expression in the batch, fed-batch, continuous culture, and solid-state in the bioreactor requires knowledge of fundamental types of kinetic rate expression. A viable inoculum, a carbon source, an energy source, nutrients, and physicochemical conditions are the prerequisites for the microorganism's growth. When all of these conditions for growth are met, the pace of expanding viable cells will be proportional to the concentration of viable biomass. Two categories of particular growth rates exist: substrate-dependent and substrate-independent.

We need to examine the estimation of products generated by microbial cells either extracellularly or intracellularly. Adenosine triphosphate, which the cell produces as chemical energy, is necessary for the microbial cell to continue its biological function.

Using carbon sources as a substrate, the cell can create ATP both anaerobically and aerobically. In the aerobic process, ATP is produced by the breakdown of carbohydrates into CO_2 and water while in the anaerobic process, ATP is produced by the breakdown of carbon sources into simple sources like ethanol, lactic acid, CO_2, water, etc. but without the presence of oxygen. The cell secretes the products back into the media. Depending on the media and other external conditions, microbial cells also release a variety of exoenzymes, polysaccharides, and antibiotics in addition to alcohol and acids. Lipids and glycogens, for instance, are formed in media that are high in carbon sources and contain few metal ions. When a system is being optimized, many characteristics and variables related to modeling must be taken into consideration. Fruits, cereals, and other items that go through fermentation often have yeast on them. The anaerobic process that breaks down the sugars in fruits and grains produces carbon dioxide, which forms bubbles that give a peculiar appearance (Demirbas, 2007). By inoculating a desired organism or a collection of organisms in the appropriate media, fermentation is accomplished.

When the media is contaminated with a bacterium and is subjected to heat or any other sterilizing treatment, death occurs at a steady rate. For instance,

90% of the microbial population perishes after receiving heat treatment for one minute. The same procedure is administered to kill the remaining population again, and 90 percent of microbes die. If a graph is created for this method, we obtain a constant curve for the death rate of microorganisms because we must repeat the treatment to eliminate the entire microbial population. The majority of sterilizing procedures work by destroying pollutants/contaminants that are present in the media being utilized. The contaminants can be eliminated by a variety of methods. Because media is important for nutrient exchange and waste elimination in the cell, the plasma membrane found inside the contaminant cell membrane is the target of sterilizing agents. The plasma membrane is disturbed by the sterilizing chemicals, which results in incorrect nutrition exchange and cell component leakage, which ultimately results in death. Bacteria make up the majority of the microorganisms that behave as contaminants. Enzymes, which are proteins that are particularly sensitive to heat treatment because they easily denature, are abundant in bacteria. Any type of sterilizing treatment, such as heating, radiation, or chemical treatment, can quickly destroy the DNA and RNA that carry the microbial genetic information.

1.9 STERILIZATION

Humans used a variety of sterilizing techniques during the early Stone Age, including drying and salting. Since ancient times, heating has been one of the best means of sterilizing.

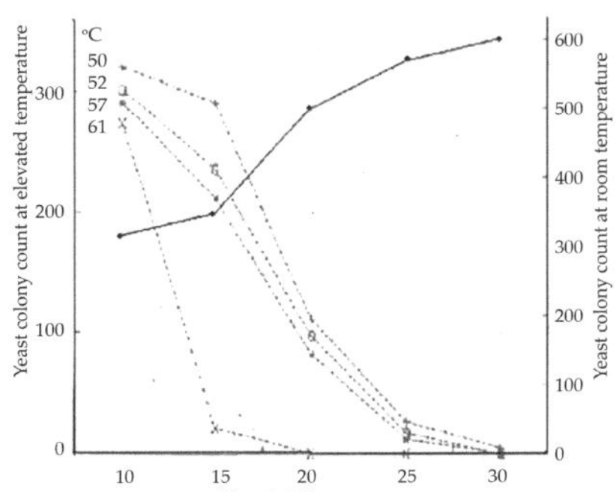

Figure 1.8. Thermal death point.

(**Source:** *https://www.researchgate.net/profile/Mohamed-Wagih-3/publication/228512448/figure/fig2/AS:670711869865997@1536921579616/Determination-of-the-Thermal-Death-Point-TDP-of-the-yeast-contaminant-when-heat-treated_W640.jpg, Creative Commons License)*

But because of the introduction of some heat-resistant microorganisms throughout time, different approaches are needed. Various physical sterilizing techniques involve the use of heat, pressure, and radiation(Gould, 1996). The use of heat is favored because it destroys microbes by denaturing the enzymes that are already present and altering the three-dimensional structure of proteins, rendering them inactive. The following three variables affect how heat sterilization works.

Thermal death point: (Figure 1.8) The temperature below which bacteria are suspended in a solution will die after 10 minutes (Slatter & Halvorson, 1947). The thermal death time is the shortest amount of time necessary to kill bacteria suspended in a solution at a specific temperature. Decimal reduction time: the amount of time, measured in minutes, needed to kill microorganisms at a specific temperature.

Dry heat sterilization and moist heat sterilization are two types of heat treatment. By burning or incinerating materials, dry heat sterilization eliminates the associated negative consequences. The method of moist heat sterilization kills microorganisms by coagulating their proteins. The most popular moist heat sterilization technique uses steam to destroy microorganisms at a high temperature and under the right pressure. Glassware, culture medium, equipment, solutions, etc. can all be sterilized with this.

The high pressure causes changes in the protein and carbohydrate molecular structures, which renders cells inactive. Radiation has different impacts on bacteria depending on its intensity, wavelength, and duration. Ionizing and nonionizing radiation are the two forms of radiation that kill microorganisms.

The wavelength of nonionizing radiation is more than 1 nm. An example of nonionizing radiation is ultraviolet light, which kills microorganisms by dimerizing thymine in their DNA. When cells reproduce, these dimmers stop DNA replication.

The microorganisms can be killed very effectively by UV light with a 260 nm wavelength (Yang et al., 2011). Glassware, air, culture media, etc. can all be sterilized with the use of these radiations. Ionizing radiation includes high-energy X-rays, gamma rays, and electron beams with wavelengths shorter than 1 nm. These radiations ionize the water and produce extremely reactive hydroxyl radicals. These radicals destroy microorganisms by rupturing DNA or RNA through a chain of chemical events.

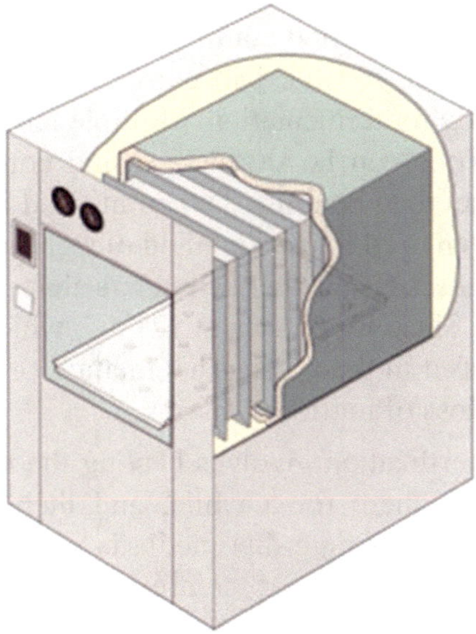

Figure 1.9. Autoclave.

(**Source:** *By Priorclave North America - Own work, CC BY-SA 4.0, https://commons.wikimedia.org/w/index.php?curid=57527103, Creative Commons License)*

1.10 HANDLING CONTAMINATION

By allowing any liquid or gas to pass through a membrane containing pores, it aids in the elimination of contaminants. To get a liquid or gas to pass through a filter that is kept on a container inside of which a vacuum is being formed, pressure must be applied. Steam sterilization can be accomplished using either regular steam produced by avoiding medium evaporation or steam under pressure used in autoclaves (Figure 1.9). It is common to utilize a Koch or Arnold steamer, which has a copper cabinet with a conical lid to aid in the drainage of steam produced and a perforated tray positioned above the water line to aid in correct steam distribution. Intermittent sterilization is a three-day method that requires a temperature of 100 °C for 20 minutes each day (Oh et al., 2009). This technique effectively eliminates bacteria but leaves anaerobic spores unharmed. The autoclave is another form of the steamer that is employed. It produces steam under pressure, and it boils water when the pressure inside the vessel is equal to the atmospheric pressure outside. This can be done in either of two ways, including heating treatment, or autoclaving. This can be carried out in a separate cooker or fermentation vessel. The media can be sterilized in cookers while the fermenters are being cleaned and ready for fermentation. Multiple fermenters can be heated in a single cooker. Time can be saved by using concentrated media for sterilization, which can then be diluted when utilized in a fermenter. Since only one cooker is often used for this method of sterilization, a complicated pipeline network connecting all the fermenters that deliver the sterilized material is necessary. Since it operates at 121 °C, which is almost possible for all nutrients included in the media, this method of sterilization ensures that there is minimal loss of nutrients.

This method of sterilization involves heating the medium to a specific temperature, holding it there for a while, and then cooling it. Holding time refers to the amount of time that media is maintained at a particular temperature. The media is heated continuously in a heat exchanger, and during the holding phase, it is maintained at that temperature in an insulated coiled tube. To cool the material to a temperature acceptable for fermentation, sequential heat exchangers are used. There are two different kinds of continuous sterilizers: one utilizes direct heat from the steam injection, while the other employs indirect heat. At regular intervals, small media increments are continually processed, and the heating and cooling periods are carefully controlled. To save heat before it enters the sterilizer, spiral heat exchangers are employed to cool the heated media with the assistance of unsterile media that are brought in for sterilization. Continuous steam injectors are used to immediately inject heat into the media, although they have certain advantages and disadvantages. While this process is highly manageable, economical, and has a very quick heating time, it also produces foam during heating, which may result in contamination. When used in combination with flash cooling, which involves creating a vacuum chamber with an expansion valve through which the media is passed, the media is instantly cooled. For starch-rich media, a mix of direct and indirect heat exchangers may occasionally be used. The most effective option for continuous sterilization is a spiral heat exchanger. A plant is constructed out of holding tubes and interconnecting spiral heat exchangers. This plant is closed-circuit sterilized with hot water before receiving an unsterile medium. The holding tube and fermenter are steam sterilized. In this process, heat is preserved by using a cool, non-sterilized medium to cool the sterilized heated media. To increase the efficiency of the fermentation process, several additives are added throughout the process. The sterilization of additives can be done in batches or continuously, depending on the biochemistry of the additives and the amount being supplied. When using a lot of additives, continuous sterilization is suggested. Steam is injected into the liquid during batch sterilization.

1.11 MEDIUM

Since the medium is cooked separately, the fermenter vessel must be sterilized before the fermentation can start. Steam is delivered into the fermentation tank by several entrances, and one valve is released to allow the steam to slowly exit the vessel, heating the coils or jackets inside. For 20 to 25 minutes, steam is introduced at a pressure of 15 psi. Sparing is used to remove sterile air from the vessel, and a positive pressure is kept in place to stop the vacuum from forming inside the vessel (Attri & Goel, 2023).

By colliding with one another, very minute particles present in the fluid are separated using Brownian motion. Small particles get captured in the filter fibers when they stray from the regular flow. The filtration of gases can be done more successfully with this method. With this technique, the suspended particles that are present in the fluid are separated based on momentum.

The fluid finds a path of least resistance through the filter, and the suspended particles, propelled by their momentum, follow it in a straight line and get restrained in the filter. The charge on the filtering membrane pulls in the oppositely charged suspended particles. The filter, which has apertures of various sizes, is made of fibers. Larger particles are excluded, and occasionally smaller particles are also unable to get through the filter membrane because their path is impeded by larger particles and they are also eliminated.

The continual input of sterile air, which can be sanitized by heat or filtration, is necessary for anaerobic fermentation. Most often, filters with fixed pores are employed to sanitize the air. Hygroscopic materials like PTFE are employed in these filters. Depth filters are made of glass wool and are used to catch airborne microorganisms. These filters function through inertia, gravitation, diffusion, electrostatic attraction, and other mechanisms. After sterilization, these glass wool filters can be simply used again. Glass fiber cartridges are currently used since glass wool filters eventually become destroyed from the action of steam. Filters in membrane cartridges are made of nylon, cellulose, or polysulfone. They are simple to operate and use.

1.12 BIOTECHNOLOGY

The term biotechnology became widely used to describe chemical engineering processes that enlist the aid of microbes in the middle of the 1970s, in particular fermenter control, design, purification, and product recovery. Food, pharmaceutical, and alcoholic beverage industries use fermentation technology on a massive scale.

The synthesis of biochemicals, biomolecules, and biofuels has utilized the fundamental fermentation principle in conjunction with cutting-edge genetic engineering techniques. Fermentation is a word used by some microbiologists to describe the process of producing bulk cultures with the aid of microorganisms. The product from a naturally occurring or genetically modified strain is referred to as biomass production and can either be the cell itself. Along with the components of the medium, the fundamental principle of fermentation is also dependent on microbial compatibility. The moment bacteria ingest the nutrients from the medium, they will create primary metabolites for their maximum growth (log phase). Secondary metabolites are produced as a result of primary metabolites (stationary phase). Therefore, the kinetics of microbial growth mirror the fermentation process. By preserving the physiochemical conditions and advancing rising microorganisms in a nutrient medium, fermentation transforms feed into desirable products. For instance, during the metabolic process of producing ethanol, yeast cells produced two ATP molecules. The by-products of a metabolic form could be further oxidized. This process involves the chemical breakdown of bioorganic molecules into simpler chemicals with the assistance of bacteria, yeast, or fungi, either separately or in combination. The behavior of this kind can be seen throughout its growth, development, senescence, and death.

When the bacterial count is increased, high-demand products such as wine, beer, cider, vinegar, ethanol, cheese, hormones, antibiotics, complete proteins, and enzymes are released into the media, and these are the by-products of their life cycles that humans can use. The foundation of fermentation is the growth of the organism in a suspended medium containing all necessary ingredients in a favorable environment. Complex organic molecules are fragmented down into simpler chemicals or nutrients during this process. Amylase converts carbohydrates, reducing starches and complex sugars into simple sugars, and lipase hydrolyzes complex fat molecules into simpler free fatty acids (Chatterjee et al., 1997). Protease, on the other hand, breaks down protein molecules first into polypeptides and peptides, followed by amino acids (Christensen et al., 2022). During the fermentation process, a few other significant by-products are also released. Fermentation is not the only mechanism to obtain energy in the absence of oxygen. It has been vital for us to ensure that only desired microorganisms begin to thrive on the substrate.

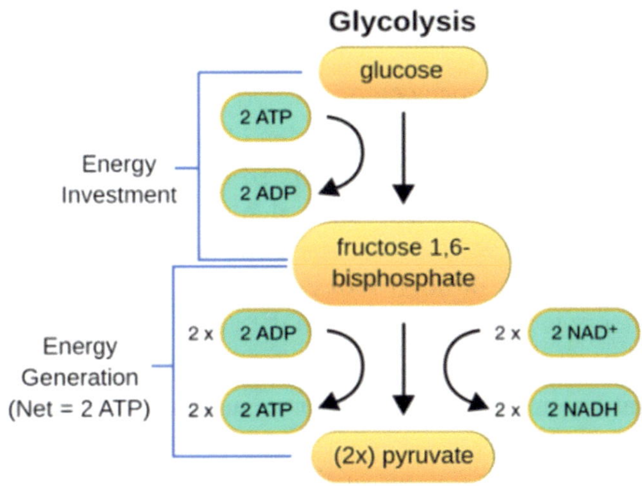

Figure 1.10. Glycolysis.

(Source: Luo Q, Ding N, Liu Y, Zhang H, Fang Y, Yin L. Metabolic Engineering of Microorganisms to Produce Pyruvate and Derived Compounds. Molecules. 2023; 28(3):1418. https://www.biologyonline.com/dictionary/glycolysis, Creative Commons License)

This procedure has made sure that other microbes are suppressed before they may ruin the fermentation process and produce an unintended and unanticipated end product, such as food poisoning or pathogenic microorganisms. Unpasteurized milk has been naturally degraded, yielding a soured product that has been employed in the baking sector to enhance the texture of the bread. Pasteurized milk, however, has been ruined and generates an unpleasant by-product that has been discarded. The primary cause of this discrepancy has been attributed to pasteurization, which altered the microbial habitat. If pasteurized milk was left out without refrigeration, unwanted microorganisms thrived and multiplied while beneficial microbes perished. Lactic acid, non-pathogenic bacteria proliferate and replicate more quickly in unpasteurized milk than any other harmful bacteria.

1.13 MECHANISM

A complementary mechanism called glycolysis (Figure 1.10) enables ATP synthesis even in the absence of oxygen. Through NADH oxidation, fermentation regenerates NAD+, which can participate in glycolysis to increase ATP production. Pyruvic acid, which is generated after glycolysis and can react with hydrogen, is created. If the normal pathway for hydrogen to reach oxygen is blocked, pyruvic acid stops functioning as a fuel and transforms into a hydrogen acceptor.

Most microorganisms make lactic acid by using pyruvic acid as a hydrogen acceptor, which is also produced by ethyl alcohol in plants and certain bacteria and mammals. Ethyl alcohol is the primary element in alcoholic beverages. Only about 5% of the energy produced by aerobic respiration is produced through fermentation (Tran & Unden, 1998). This amount of energy is sufficient to sustain the existence of bacteria, yeast, and other anaerobes.

Fermentation energy is insufficient to support these species. They can increase their aerobic energy through fermentation. The changes that occur during fermentation are a result of the enzymatic activity. Enzymes have controlled biochemical reactions with precision. Enzymes (Figure 1.11) have been recognized as catalysts because they orchestrate reactions rather than being used in them. They vary in moisture content, temperature, pH, substrate and inhibitor concentrations, and ionic strength in nature and are proteinaceous. Each enzyme has specific requirements that must be met for it to work well. Temperature and pH extremes will denature the protein and reduce enzyme activity. Because of their extreme sensitivity, it is easy to regulate their enzymatic reactions by making small changes to the temperature, pH, moisture content, or other reaction conditions.

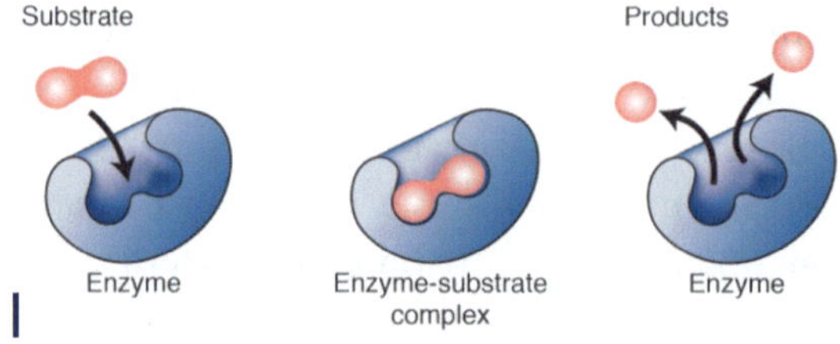

Figure 1.11. Enzymes.

*(**Source:** https://www.genome.gov/sites/default/files/tg/en/illustration/enzyme.jpg, Creative Commons License)*

1.14 FOOD SECTOR

In the food sector, enzymes have a variety of activities, including sugar conversion, protein modification, starch liquefaction and saccharification. The significance of microbial enzymes in the fermentation of fruits and vegetables is significant.

Making sure that just the desired yeasts, molds, or bacteria begin to multiply and thrive on the substrate is crucial for any fermentation. The fermenting process takes place in a container called a fermenter or bioreactor. The design and nature of the fermenter vary depending on the sort of fermentation performed.

Every fermenter has sensors to measure several fermentation parameters, including liquid, temperature, pressure, pH, and concentration. Methanol is used as the substrate in the externally recycled airlift fermenter to create bacterial biomass.

Every country has its distinct fermented food varieties that represent the main cuisine and raw materials that can be found there. The advancement of fermentation technology is significant under anaerobic situations when oxidative phosphorylation is not present to maintain the generation of ATP (Adenosine triphosphate) by glycolysis. Lactic acid is made from pyruvate by a process known as homolactic fermentation. Alcoholic fermentation is the process of turning pyruvate into ethanol and carbon dioxide(Zamora, 2009). Heterolactic fermentation is the process of producing lactic acid as well as other acids and alcohols. Pyruvate is transformed into a variety of chemicals during fermentation. Acetone and butyric acid, for example, can be created through fermentation. Lactic acid and ethanol are two examples of fermentation products.

An organic substance is partially oxidized during the metabolic process of fermentation employing intermediates as either electron donors or electron acceptors. When the substrate is oxidized, phosphorylation at the substrate level occurs, and ATP is produced directly from the process. The process is accompanied by the formation of two molecules of ATP, which are formed as a result of the conversion of glucose into two molecules of lactate.

Like fermented foods, fermented products are extremely important because they preserve and make available vast amounts of nutritious foods in a wide range of textures, tastes, and fragrances that enhance the nutritional value of food for human consumption. They have provided us with vinegar, alcoholic foods, beverages, cheeses, yogurts, sauces and pastes with meaty flavors, as well as sausages, and pickled vegetables. While wealthy nations employ manufactured foods fortified with vitamins, developing nations have relied on biological systems that are enhanced with vitamins and critical amino acids.

The uses of fermentation are expanding quickly beyond basic foods like cheese, wine, beer, and bread to include high-end goods, food additives, and pharmaceutical substances. Due to the high price of gasoline and the deterioration in the use of fossil fuels, fermentation techniques have shown to be both affordable and simple. By creating cost-effective bioethanol using fermentative processes, biofuels have attracted increased attention. Fermentation technology is one of the oldest processes still in use today in the biotechnology industry.

The fermentation process is at the core of fermentation technology and is traditionally used to prepare wine, bread, idli, cheese, sausages, and other food products.

To put it another way, fermentation is the biological process triggered by microorganisms that result in the production of heat and carbon dioxide. Depending on the amount of product, fermentation has been done in small or big fermenters or bioreactors. The names fermenters and bioreactors are similar yet serve different functions. A form of closed-vessel bioreactor called a fermenter provides a regulated and sterile environment for the growth of microbes in a liquid media to produce a variety of chemicals.

1.15 FERMENTERS

Fermenters are a kind of closed system used to perform biological processes in a regulated environment (Pulz, 2001). Fermenters are used as a receptacle for performing chemical procedures involving microorganisms and/or chemicals generated from bacteria. By providing oxygen and nutrients to cells, fermenters typically regulate the metabolic waste products produced by the cells and preserve both a biomechanical and a biochemical environment. Recombinant products like hormones and vaccines, product biotransformation i.e., L-sorbitol, steroid biotransformation, and enzyme manufacturing can all be done in fermenters. These internal reactions could be either aerobic or anaerobic. These fermenters come in sizes ranging from liters to cubic meters and are composed of stainless steel. To accommodate a wide range of substrate products and biocatalysts, many bioreactor designs have been created. Depending on the type of microorganisms grown in these fermenters, they may be submerged, attached to the surface of a solid medium or both. Since all microorganisms do not require particular attachment surfaces, a wide range of microorganisms can be employed in suspension fermenters, which can be handled at a much higher level than immobilized fermenters. Immobilization is the process of entangling or attaching a wide range of cells or particles to an immobilized substance. In contrast, the effluent will be used in a continuously operating process to remove the microorganism from the fermenter.

Because they support and regulate living things, bioreactors are distinct from traditional chemical reactors. The microorganisms should be able to demonstrate their activity in the bioreactor under specific conditions. A fermenter is a form of bioreactor that uses a living cell as the biocatalyst, and the terms fermenter and bioreactor are frequently used interchangeably. The sizes of the bioreactor can range from the microbiological cell (a few mm3) to pilot level fermenters (0.3–10 m3), plant scale fermenters (2–500 m3), shaking flask (100–1000 ml), and fermenters of laboratory scale (1–50 l) for applications in the production of large volume industrial products (Witzig et al., 2002). It is necessary to closely monitor the conditions of bioreactors, including flow rates, gas levels (air, oxygen, nitrogen, and carbon dioxide), pH, temperature, foam generation, agitation speed or circulation rate, etc. Although the fermentation process has been known for thousands of years, Louis Pasteur, a French scientist, did the first scientific study on it in the 1850s using the formation of lactic acid. The advancements in fermentation techniques have been essential in the development of fermenters and bioreactors. In the past, the fermentation process was used to brew and produce alcoholic beverages, but as our understanding of fungi and bacteria has grown, more effective fermenters have been created. Fermenters have been used for the regulated development and maintenance of bacterial and fungal cells, whereas bioreactors have been utilized for the growth and maintenance of mammalian and insect cells.

1.16 MICROORGANISMS

Microorganisms can complete their task in bioreactors under ideal conditions with few contaminants. However, environmental factors including nutrient concentrations, temperature, dissolved gas concentrations, and pH inside the bioreactor have traditionally had an impact on the growth of microorganisms. To maintain the temperature of the fermentation medium, cooling jackets, coils, or both may be utilized. External heat exchangers may also be used in exothermic fermentations. The optimal conditions needed for the selected product, as well as the production size, quality, and quantity of microorganisms, determine the bioreactor design and mode of operation.

Nutrients are continuously added at the start of fermentation in a fed-batch fermenter. Small amounts of acid or base have been used to modify and maintain the pH depending on the fermentation process. Since reactant gases, particularly oxygen, must be introduced to aerobic fermenters, air or filtered oxygen must be continuously added as oxygen is relatively insoluble in water.

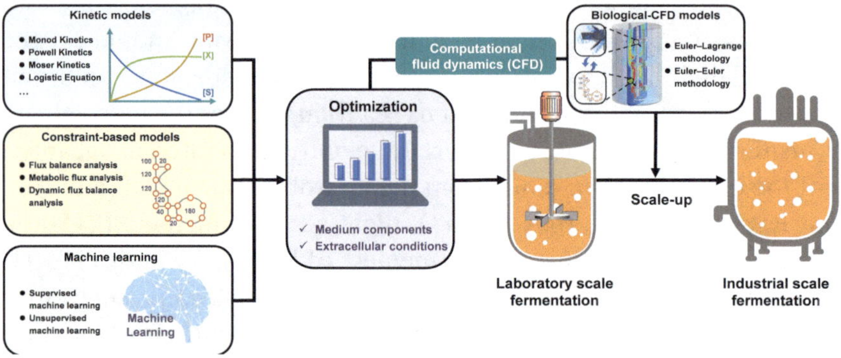

Figure 1.12. Design of a fermenter.

(Source: Du Y-H, Wang M-Y, Yang L-H, Tong L-L, Guo D-S, Ji X-J. Optimization and Scale-Up of Fermentation Processes Driven by Models. Bioengineering. 2022; 9(9):473. https://doi.org/10.3390/bioengineering9090473, Creative Commons License)

The bubbles aid in mixing the fermentation medium and remove gases like carbon dioxide. The transfer of the ideal oxygen level is the rate-limiting phase in aerobic fermentation. To mix the nutrients and maintain homogeneity in the fermentation, agitation can occasionally aid in oxygen transport. To meet the design criteria, factors such as sterilization, process control devices, ease of construction and measurement, scale-up, regulating strategies, flexibility and compatibility with upstream and downstream processes, antifoaming methods, etc. are crucial.

1.17 DESIGN

The design of a fermenter (Figure 1.12) should allow for good drainage, proper aeration and agitation, harvesting of the products, and sterile conditions that can be built inside of it. Before designing a fermenter, one should also consider the following additional factors. The vessel needs the least amount of labor and the work can be done aseptically under controlled circumstances. The internal surface should be smooth and composed of inexpensive materials that produce outstanding results. There are two different types of fermenter vessels utilized; glass is favored for small-scale use while stainless steel is used for industrial purposes. Glass is resistant to corrosion. The internal reaction is simple to observe. Through the use of an autoclave, sterilization is accomplished. These tiny fermenters have a diameter of around 60 cm. Large-scale fermentations are typically conducted in stainless steel. These vessels may be able to withstand corrosion and pressure and in situ sterilization takes place. The microbial activity and agitation in the fermenter vessel result in the production of heat. Heat is either added to the system or removed from it to maintain the temperature inside the vessel. Internal coils or thermostatically regulated baths are typically utilized to generate heat, while silicone jackets are used to dissipate excess heat. It includes two silicone mats that are bonded together with heating wires. Heat removal becomes laborious if the size is exceeded, covering the surface with the jacket, and then cold water must be cycled inside the internal coils to maintain the precise temperature.

To ensure adequate agitation and a longer aseptic operation, the stirrer shaft is sealed using a sealing assembly. The fermenter has different types of sealing assemblies. In the case of a packed gland seal, the gland presses against the asbestos packing rings that have been used to seal the shaft. Packing rings have been inspected and replaced regularly to prevent inadequate heat penetration. Spring pressure is used to push two pieces together. The steam condensate is utilized to cool and lubricate the seals during operation. The driving magnet will be connected to the drive shaft and seized on the exterior of the head plate in a bearing. Another is that the driven magnet will get seized in bearings on the interior of the head plate, at the end of the impeller shaft. Baffles are made of metal strips that are radially fastened to the wall to prevent vortices from expanding the aeration capacity. The microbial growth on the fermenter sides has been reduced using baffles. Through agitation, impellers provide homogenous microbial cell suspension in homogeneous nutrient media. In the suspension culture, impellers combine the bulk fluid, solid particles, and gas phases.

Disc turbines are devices having a disc with numerous rectangular vanes on it. They allow air from the sparger to strike the disc's bottom and move it in the direction of the vanes, where it is broken up into smaller air bubbles. Open turbines with variable pitch also include a vanned disc attached to the agitator shaft along with a pair of propeller blades. This turbine's air bubbles disperse without touching any surfaces. Sparger ensures that the vessel is properly aerated so that the microorganisms have enough oxygen for their metabolic processes. The porous sparger is employed in non-agitated vessels of laboratory size and is constructed of ceramic or sintered glass. The nozzle sparger features a single, partially open pipe. Because they don't get clogged and offer less pressure, this form of sparger is frequently used. In a combined sparger agitator, the air is introduced through a hollow agitator shaft, and it is then released from holes in a drilled disc to connect to the main shaft base. A baffled vessel receives adequate aeration when the agitator is run at a variety of rpm. Silicone tubes serve as feed ports. In the fermenter, they are utilized to add nutrients and acid or alkali (Stratton et al., 1998).

Foam sensing and control devices work together to control foam. A probe has been put through the fermenter's top and positioned at a specific height above the broth's surface. A current will flow through the circuit once the foam level rises and reaches the probe tip. The pump will be activated by this current, and antifoam will be released right away to deal with the problem. In the fermenter, valves are utilized to regulate the flow of liquid. Globe valves, butterfly valves, ball valves, and diaphragm valves are the common types of valves utilized. Globe valves can be used for a variety of applications; however, they cannot control flow. Butterfly valves are used for large-diameter pipelines that run at low pressure yet are not ideal for aseptic conditions. Ball valves are appropriate for an aseptic environment. They deal with mycelial broths and work in hot environments. Flow adjustments are made easier using diaphragm valves. The air and pipe layout includes a safety valve that can work under pressure. These valves assist in keeping the pressure within acceptable ranges. With the aid of microbes such as bacteria, fungi, mammalian cells, and plant cell systems in the form of initiator cultures, various biochemical processes are carried out in fermenters and bioreactors to produce different biological products. They offer a favorable environment for the synthesis of metabolites. Traditional goods like wine, cheese, and beer are produced in non-stirred, non-aerated bioreactors. Stirred aerated reactors, however, are based on contemporary technology.

The fermenter types that are utilized are the fluidized bed, packed bed, photobioreactor, bubble column fermenter, membrane fermenter, and continuous stirred tank. The flow rate of the fermenter can be changed to achieve a suitable and consistent value while maintaining optimal quantities of microorganisms, substrates, and products. The activated sludge process, which is utilized in the wastewater treatment sector, is the most extensively used continuous process. It is based on the Continuous Stirred Tank Fermenter (CSTF) (Mohammadi et al., 2012). Stirred tank bioreactors have a lot of advantages, including the ability to run continuously in the fermenter, simple temperature control, low construction costs, low labor costs, and simple operation and cleaning. For gas-liquid or gas-liquid-solid contacting devices, airlift reactors are typically used. They circulate different fluids in a clear cyclical pattern through built-in conduits. The substance inside channels is agitated by a stream of air or other gases. The gas stream aids in the material exchange between the medium and the gas phase; oxygen is frequently transported to the liquid. When the gas phase is introduced, products that have resulted from reactions are ejected. Airlift reactors come in two primary varieties: internal-loop vessels, which have baffles arranged in a single vessel to provide circulation, and external-loop vessels, which provide circulation through separate and independent channels. The pattern of both types of vessels can be changed to provide different fluid dynamics.

Many sectors of the chemical, petrochemical, and biochemical industries employ bubble column reactors. These reactors are inexpensive to operate, have a simple design, and require little maintenance. They have a cylindrical shape and a height-to-diameter ratio of 4:6, and air or gas is injected into them at the foot of the column via perforated pipes, plates, or metal microporous spargers. Accurate air or gas flow rate maintenance is performed to ensure proper O_2 transfer or mixing. To increase the reactor's efficiency, perforated plates are attached to the fermenter. In many chemical processing applications, such as absorption, distillation, separation processes, and catalytic reactions, packed bed reactors, also known as fixed bed reactors, are used. It comprises a partition-like tube or channel with catalyst pellets or particles on it, through which liquid passes. This reactor has many benefits, including a high catalyst conversion rate, ease of management and construction, more efficient contact between the reactant and catalyst than in other types of reactors, increased product formation as a result of the increased reactant/catalyst contact, low construction, operation, and maintenance costs, and the ability to function effectively at high temperatures and pressures.

CHAPTER

2 Food Fermentation

Contents

2.1 INTRODUCTION

In rocks that are 3.3 to 3.5 billion years old, fossilized bacteria have been discovered (Altermann et al., 2006). They were the first life forms to emerge on the planet. They were probably bluc-grccn algac, (Figure 2.1) which have a pigment that allows them to produce carbohydrates from solar light. Similar to all other current forms of life, they include deoxyribonucleic acid (DNA) and ribonucleic acid (RNA).

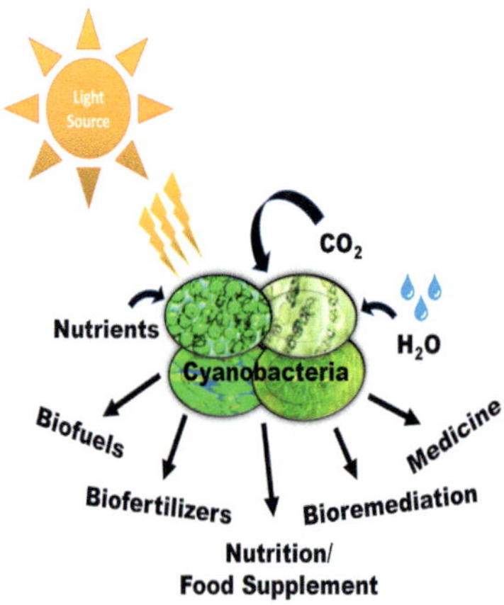

Figure 2.1. Blue-green algae.

(**Source:** *Zahra Z, Choo DH, Lee H, Parveen A. Cyanobacteria: Review of Current Potentials and Applications. Environments. 2020; 7(2):13. https://doi.org/10.3390/environments7020013, Creative Commons License)*

They include the lipases, proteases, amylases, and other enzymes needed to hydrolyze proteins, starches, and lipids for recycling. Without bacteria, the planet Earth would be a gigantic landfill for all kinds of organic waste. At first, bacteria used organic matter as food to fuel their metabolic processes. Microorganisms later absorbed dead plants and animals as plants and animals evolved, and they later developed the ability to infiltrate plants and animals, causing disease. The plants served as the food source for animals. By creating a lignocellulosic framework that resists microbial invasion, plants were able to do this. All the chemical and enzyme reactions required for food fermentations were already in place millions of years before humans first appeared on the planet. These reactions included the fermentation of fruits and fruit juices to produce wine and vinegar, the germination of grains to produce alcoholic beverages, and the souring of milk. Food became spoilt and people learned to stay away from it when bacteria developed displeasing odors or scents in it or toxins that led to disease or death. Humans learned to value and crave certain foods if the invasion of the food components by microorganisms resulted in appetizing scents and flavors. These marked the beginning of fermented foods, which now include lactic acid products like sauerkraut, sour milk, cheeses, wines and beers, vinegar, and hundreds of other items.

2.2 SALTS

Salt, often known as table salt, sodium chloride, or sea salt is a combination of sodium, potassium, magnesium, and calcium salts found in seawater. In ancient times, soldiers earned some of their pay in the form of salt (or salary). When present in the right amounts, salt stops putrefaction and promotes regulated protein hydrolysis. The ability of salt to preserve fish or other animal tissues was also found relatively early by humans, especially when combined with sun drying. Today, fish that has been salted and sun dried (Figure 2.2) is a common diet in many regions of the world (Belton et al., 2022).

Figure 2.2. Sun-dried fish.
(**Source:** https://upload.wikimedia.org/wikipedia/commons/7/7a/Dry_fish_stand_near_Banda_Aceh_%288035448940%29.jpg, Creative Commons License)

Due to the likelihood that materials held in seawater would also experience lactic acid fermentation, seawater may have also contributed to the early lactic acid fermentation and preservation of plant materials. Anthropologists have proposed that human settlement and the pursuit of agriculture were driven by alcoholic fermentation and the craving for alcohol. Without fermented foods, humans would not have been able to survive for thousands of years. Foods are preserved through fermentation, which also increases digestibility and enriches substrates with crucial vitamins, amino acids, and fatty acids.

Some of the grains stored at the beginning of agriculture occasionally became soaked by rain, germinated, and fermented to make beers. The fermentation of agave pulp juices yields pulque, the oldest known alcoholic beverage in Mexico. The dextrans that Leuconostoc mesenteroides generate provide the beverage texture (Juvonen et al., 2015). Saccharomyces cerevisiae, a yeast, or Zymomonasmobilis, an alcohol-producing bacteria, are the sources of ethanol. In Mexico, pulque, which is extremely high in B vitamins, is crucial for the nutrition of those who are less fortunate economically. Acetic acid is present in alcoholic beverages, but the amount is usually little as long as the fermentation is anaerobic. By depositing a layer of CO_2 on the surface of the fermenting materials, carbon dioxide production supports the maintenance of anaerobiosis. A percentage of the ethanol is transformed to vinegar when the alcoholic fermentation ceases due to Acetobacter sp., which becomes active as soon as oxygen is present. Due to its germicidal properties, vinegar is a very valuable pickling agent and can even be used medicinally. When seeds fall to the ground, they either germinate and grow into new plants or, as the first stage of recycling, they become dominated by bacteria. Since the seed coat is rather resistant to microbial growth, molds that generate proteases, lipases, and amylases that hydrolyze the different parts of the seed are frequently the first compounds to enter the cotyledons. The mold-covered seeds then serve as a source of enzymes. Such moldy seeds, known scientifically as "koji," are useful for hydrolyzing the proteins, lipids, and carbohydrates in other plant or animal-based products (Hachmeister & Fung, 1993). Salt, pork, fish, or poultry were combined with millet koji, which was then kept in a bottle for a hundred days.

2.3 SOYBEANS

Soybeans are a food source because they are high in protein. They need to be cooked and hydrated/soaked to be tasty. Soybeans are not prone to microbial deterioration as long as they are kept dry. As is the case with most seeds, they are vulnerable to bacterial and mold infestation after being soaked. Soybeans are soaked, cooked, covered with rice straw, and placed beneath the ceiling of the house where it is warm to prepare soybean paste. The straw-borne Aspergillus molds outgrow the soybeans in around 30 days (Park et al., 2003). Then, sea salt brine is added to the mold-covered soybeans, and they are left to break down for a year or longer. Proteins, lipids, and carbohydrates are broken down by mold enzymes, resulting in tasty soybean paste with peptide and amino acid flavors. The liquid that is produced from the soybean paste is a delicious soy sauce similar to tamari that is particularly rich in amino acids/peptides. Given that the initial stage of Indonesian tempeh fermentation involves growing soybeans with a mold called Rhizopus oligosporous or a comparable strain, it is closely related to the fermentation of soy sauce. The soybean cotyledons (Figure 2.3) are mixed with the fungus mycelium to form a compact cake that may be deep-fried or chopped into cubes for soups. People in Indonesia have been performing this fermentation for hundreds of years without any formal training in microbiology or chemistry, yet they can create tempeh of the highest caliber.

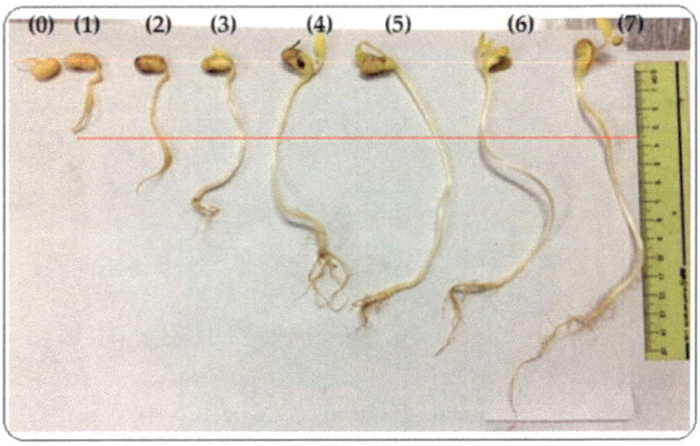

Figure 2.3. Soybean cotyledons.

(**Source:** *Le XT, Lan Vi VL, Toan TQ, Bach LG, Truc TT, Hai Ha PT. Extraction Process of Polyphenols from Soybean (Glycine max L.) Sprouts: Optimization and Evaluation of Antioxidant Activity. Processes. 2019; 7(8):489. https://doi.org/10.3390/pr7080489, Creative Commons License)*

2.4 ADVANCEMENTS

The development of a new high-tech sector to produce meat alternatives in recent years is the most unexpected aspect of tempeh fermentation. There are two main approaches. First, soybean protein is extracted and spun into fibers by putting the protein strands in a chemical solution. The resultant fibers are flavored with meat and have a feel that resembles flesh. As vegetarian meat alternatives, the dehydrated chunks are added to soups and other culinary products. It is an extremely complicated and pricey method of food preparation. Indonesian tempeh accomplishes a similar goal through fermentation, wherein mold mycelium gives a texture resembling that of meat, and the resulting goods are affordable to the typical Indonesian. Growing edible strains of the Fusarium graminearummold mycelium, extracting the mycelium by centrifugation/filtration, adding meat tastes, and then dehydrating the mycelium are steps in making meat substitutes that are even more similar to the tempeh process. Compared to the tempeh method, this technology is more sophisticated and expensive. By allowing Rhizopus oligosporus (Figure 2.4) mycelium to cook dehulled soybean cotyledons, Indonesian tempeh can mimic the texture of meat substitutes (Jurus & Sundberg, 1976).

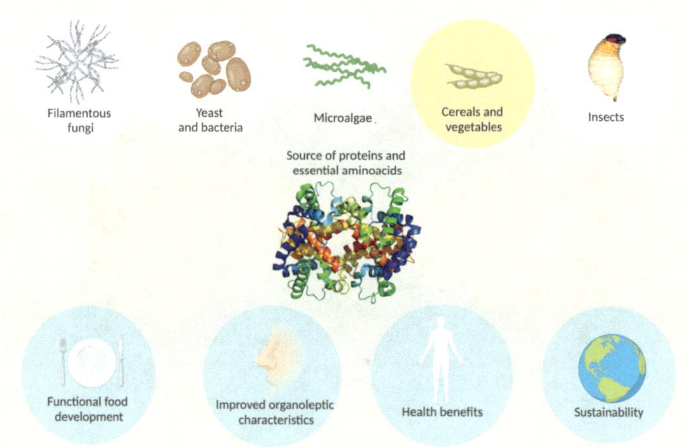

Figure 2.4. Alternative to Meat: State of the Art and Involvement of Fermentation.

(**Source**: *Molfetta M, Morais EG, Barreira L, Bruno GL, Porcelli F, Dugat-Bony E, Bonnarme P, Minervini F. Protein Sources Alternative to Meat: State of the Art and Involvement of Fermentation. Foods. 2022; 11(14):2065. https://doi.org/10.3390/foods11142065, Creative Commons License)*

Since they helped the human race survive and flourish, lactic acid fermentations are among the oldest and most important fermentations in the world. Humans first learned about lactic acid fermentations when they began raising domesticated cows, sheep, and goats for milk. The stomachs of slaughtered animals were among the oldest containers used by humans to preserve milk. Because milk naturally contains lactic acid bacteria, it ages relatively quickly. After humans began to settle and practice agriculture, sour milk was one of the first foods to be fermented, and it is still consumed today in the form of yogurt. When soured milk was kept in animal stomachs, other lactic microbes in the environment caused the milk to curdle, lose its whey, and turn into rudimentary cheeses. Cheese types have been a staple of the human diet for ages and they still are now. Energy-efficient lactic acid fermentations often don't require any heating or cooking before or after the fermentation process. The fermentation of sauerkraut is a classic instance of lactic acid vegetable fermentation. Shredded fresh cabbage is combined with 2.25 percent salt (Cheigh, Park, & Lee, 1994). The salted cabbage is covered with a plastic top or airtight covering. Leuconostocmesenteroidesgrowth marks the beginning of natural fermentation. Lactic acid and carbon dioxide produced by L. mesenteroides wash out any remaining oxygen, assisting in the maintenance of anaerobic conditions. Lactobacillus brevis, which generates extra lactic acid and carbon dioxide, is the second organism to emerge. Lactobacillus plantarum, which generates more acid, comes next. Pediococcus cerevisiae, the final organism to emerge and produce more acid, is the last organism to develop. As long as the product is stored anaerobically, the finished product has an acidity of roughly 1.7 percent to 2.3 percent acid (as lactic acid). The sauerkraut can be consumed hot or cold, raw or cooked, as a salad.

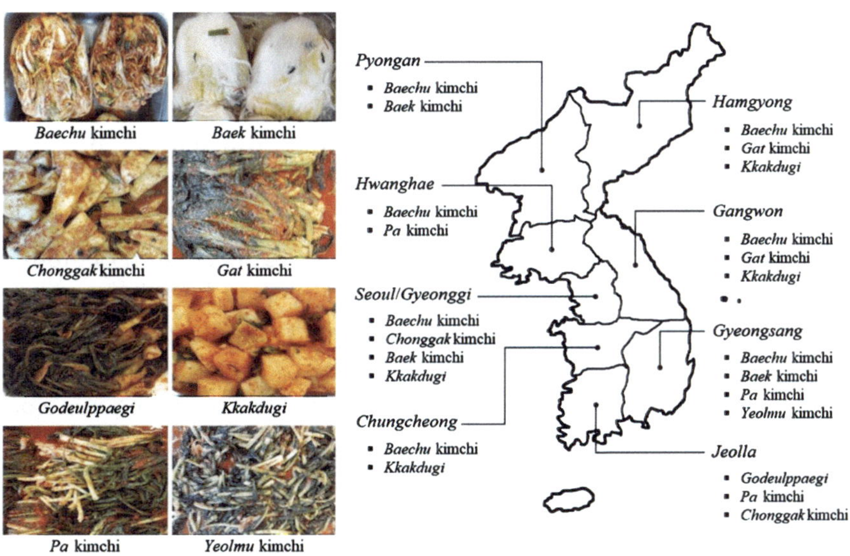

Figure 2.5. Korean kimchi.

(**Source:** *Park YK, Lee JH, Mah J-H. Occurrence and Reduction of Biogenic Amines in Kimchi and Korean Fermented Seafood Products. Foods. 2019; 8(11):547. https://doi.org/10.3390/foods8110547, Creative Commons License)*

2.5 CABBAGE

Chinese cabbage is the main ingredient in Korean kimchi (Figure 2.5) but radishes, red peppers, and other vegetables can also be used (Haque et al., 2015). The veggies are chopped, then drained and rinsed after being submerged in a strong salt brine (5-7% salt) for 12 hours-. The duration of the subsequent fermentation varies depending on the temperature. Kimchi is carbonated and has a lower acidity than sauerkraut. The majority of vegetables, including cucumbers and carrots, as well as green fruits, like limes and olives, have been subjected to lactic acid fermentation. Worldwide, lactic acid fermentation is a key technique for preserving fresh vegetables. The use of fire for heating and grinding starchy grains like barley, wheat, millet, rye, and sorghum to make them easier to eat is thought to have been the beginning of bread fermentations. When such flour is mixed with water, lactic acid organisms and yeasts in the environment start to ferment the mixture. One or more yeast species and lactic acid species are typically present in the final product. Bread has been the main food in human meals since at least 5000 B.C. Gluten, a component of wheat flour, traps the carbon dioxide released during the symbiotic heterofermentation of lactic acid bacteria and yeast (Raman et al., 2022). Perishable food products have been preserved for generations through the use of food fermentation. Fruits, cereals, honey, vegetables, milk, meat, and fish are a few of the diverse types of raw materials that are historically used for fermentation. Wine, beer, vinegar, bread, soy sauce, sauerkraut, kimchi, pickled olives, fermented milk products, a wide number of cheeses, and a variety of sausages are all considered to be fermented items. Some procedures need an inoculum, and this requirement was met by retaining a sample from the prior production. Inoculation was not required for other processes because, given the right conditions, naturally occurring bacteria in the raw materials might serve as a reliable source of the microbial flora. This is true for the manufacture of wine, sauerkraut, some fermented sausages, and raw milk cheeses.

Understanding and controlling food fermentations became possible with the discovery of microbes. In the 19th century, techniques for isolating and purifying microbial cultures were developed. Before being inoculated with specific cultures, the raw materials were sterilized or pasteurized, which made it possible to control the fermentation processes with minimum variance. By the 19th century, the employment of specified cultures had become the industrial norm in breweries. The wine, dairy, and meat sectors likewise changed their production processes over the 20th century to use starting cultures. The quality of the fermented food items significantly improved as a result of the application of microbiology and process technologies. Fermentations of food have used a wide range of microorganisms. Baker's yeast, brewer's yeast, wine fermentation starter cultures, food and feed additives, and flavoring cultures for dairy and meat products are all forms of Saccharomyces cerevisiae. The amount of annual production of baker's yeast is roughly 1 million tonnes, which is almost two orders of magnitude more than the total annual production of all other microorganisms. Yeast is generated in large quantities, which makes it a completely distinct commodity from the high-value, low-volume yeast strains created for inoculating wine fermentations and from all the other inoculants for food fermentations. The most common lactic acid bacterium is Lactococcus lactis. L. lactis is employed to some extent in the manufacture of meat, bread, and vegetable foods as well as for the production of cheese, butter, buttermilk, and other fermented milk types (Yeh et al., 2009). The second most common lactic acid bacteria are Streptococcus thermophilus. It is used in the dairy industry to make cheeses, most notably mozzarella and pizza cheese, as well as yogurt and a range of other fermented milk products. S. thermophilus is frequently employed in conjunction with other lactic acid bacteria to speed up the fermentation process because of its high acidification activity.

2.6 LACTOBACILLUS

L. delbrueckii and L. acidophilus are two Lactobacillus species that are utilized in reasonably high quantities, particularly in dairy products. L. acidophilus is also used in many probiotic goods. In fermented foods and nutritional supplements, several Bifidobacterium species are employed as probiotic cultures. Because there are so many microorganisms with largely consistent cell and colony morphologies, it is challenging to identify and classify them. Classical microbiology frequently employs phenotypes of the +/- type. Cell morphology, Gram staining, growth on different carbohydrates, gas formation, acid production, temperature profile for growth, salt tolerance, amino acid requirements, vitamin requirements, oxygen requirement/sensitivity, hemolysis, and hydrolysis of polysaccharides, proteins, and lipids are the traits/features that are traditionally used for identification.

Methods for differentiation at the strain level are based on serological tests, phage typing, or biochemical studies. In the microbiology lab, all of these conventional tests are still carried out and are crucial, especially those that directly relate to the metabolisms utilized for fermentation. However, due to the adoption of techniques from molecular biology over the past 20 years, the identification, categorization, and taxonomy of microbes have experienced significant modifications. Protein synthesis is one of the fundamental procedures that every living thing shares with others. Taxonomists can determine the evolutionary distances of organisms by comparing their degree of resemblance to the prevalence of ribosomes. The RNA molecules of the ribosome are especially useful for this purpose because of practical considerations. Lactic acid bacteria species do not have a close phylogenetic link, but they do share a common method for surviving in nutrient-rich settings.

The few primary metabolic pathways that are used by the food fermentation processes are

1. Alcoholic fermentation carried out anaerobically, which turns carbohydrates into alcohol and CO_2,
2. Lactic fermentation produces lactic acid by converting carbohydrates,
3. Fermentation of acetic acid by aerobic bacteria, which turns alcohol into acetic acid. Alcohol and organic acids, which are the principal end products, are primarily responsible for the preservation and greatly enhancing the flavor of fermented foods.

However, secondary metabolic pathways do provide a wide range of tastes and textures, producing a variety of products. Microorganisms can create hydrolytic enzymes including proteases, peptidases, lipases, glucanases, and amylases to meet their nutrient needs (Saxena et al., 2001). These enzymes can produce or break down compounds with strong flavors in addition to meeting the needs of the microorganisms. When it comes to ripened cheese and meat products, enzymatic interactions are crucial. The diversity of starting cultures is mostly due to the wide variation in the enzyme-substrate specificity among microbial species and strains. The diversity of cultures is also influenced by the accumulation profile of metabolites such as diacetyl, acetaldehyde, acetate, and formate. The precise metabolic pathway from the glucose to the primary end products, as well as the use of alternate electron acceptors, affect the balance between these metabolites. The small variations in the metabolic pathways of the cultures affect not only the production of flavor but also that of gas and texture.

The lag phase of the starter must be reduced, the growth rate must be increased, and the metabolic activity of the cells must be increased to optimize fermentation speed. Selecting strains with low sensitivity to environmental influences that may be present in fermentation can help ensure reliability.

The starting culture attacked by bacteriophages is a common problem in large-scale fermentations. The resistance of Lactococcus lactis from phage attacks has received attention. Most food fermentations produce an overall preservation effect by building up organic acids and alcohols while simultaneously decreasing the pH, reducing free sugar levels, and depleting oxygen. It has been discovered that most cultures that produce antibacterial bacteriocins have substantially stronger preservation properties. Reuterin, an antibacterial compound produced by the lactobacillus reuteri, is a low molecular weight metabolite called 3-hydroxy propionaldehyde (Soltani et al., 2021). Nisin, a bacteriocin, was found roughly 70 years ago. Lactococcus lactis strains produce nisin, a short peptide with unique amino acids as a result of post-translational changes (Beasley & Saris, 2004). More than 50 years of use as a food preservative have gone into making nisin and most nations now approve of its use. Based on structural differences, a large number of bacteriocins from lactic acid bacteria have been identified and divided into three classes. The ability of the bacteriocins to create pores in the target bacterial membrane serves as a common mechanism of action. Microorganisms isolated from fermented foods frequently possess the ability to produce bacteriocins.

2.7 HUMAN HEALTH

Metchnikoff described the lactic acid bacteria's positive impact on human health approximately a century ago. These beneficial health effects, which were later considered the probiotic effect, have been supported by research. Probiotics are now defined as living microorganisms that, when taken in sufficient proportions, give a health effect on the host. A very intricate ecology of different bacterial species makes up the bacterial flora of the digestive tract. For the digestive system to operate normally, intestinal microbial flora is required. The preservation of healthy bacterial flora is preferred since the elimination of the flora causes diarrhea or constipation. Acid and bile stability, pathogenic microorganisms, safety during usage, and clinical proof of the health effects are the primary prerequisites. A crucial component of the production of starting cultures is the industrial-scale proliferation of microorganisms. Industrial scale can range from laboratory propagation in flasks and agar plates to fermenters that are hundreds of cubic meters in size, depending on the product. The majority of cultures are formed via liquid-submerged fermentation, except a small number of mold cultures produced by sporulation on solid media. Yeast and aerobic flavor cultures are formed by aerobic fermentations, whereas the vast majority of bacterial cultures are created by anaerobic fermentations. The upstream procedures that guarantee the purity of the generated microorganisms are the most crucial.

The steps used after fermentation are intended to maximize cell density, protect the microbes, and package the finished products in a way that makes them easy to store, distribute, and utilize. By using ultrafiltration or centrifugation, the cell density can be raised. The product can be packaged in liquid, dry, or frozen form depending on the desired format and the fragility of the microbe. Microorganisms must be shielded from excessive heat, moisture, and light by the packaging material. The actual sensitivity to these variables can fluctuate not just across different sorts of cultures but also between various formulations of the same culture. Both a bulk starter cultivated locally and a highly concentrated starter culture obtained from the supplier can be used to inoculate food fermentations. The amount of various fermented foods produced in the same facility, the level of automation, the microbiology expertise, and, lastly, the economics will all have an impact on the decision between the process types. Direct inoculation of the culture results in the highest level of security and adaptability. There are options for several culture formulations in addition to direct inoculation or bulk starter. In most cases, baker's yeast is added directly to the dough without first growing in the bakery. Both fresh and dry yeast are available; fresh yeast is available as liquid, compressed, or powdered yeast, and dry yeast is available as active yeast. The most common format for mold cultures used as direct inoculants is a dry spore preparation.

2.8 BACTERIAL CULTURES

The three main forms of bacterial cultures for sale are liquid, lyophilized, and frozen. Direct cultures will need a regular supply of liquid bacterial cultures because they typically lose their activity within days. For months or even years, cultures that have been lyophilized or frozen maintain high activity, making these forms the best choice for the widespread distribution of direct inoculants. The media to be employed for the propagation is accessible from the same source as the cultures used in these systems, which are supplied by a starting manufacturer in frozen or freeze-dried form. The primary justification for purchasing starter cultures from a commercial supplier rather than growing local supplies of cultures is safety. Commercial suppliers have the technical and scientific skills required to preserve purity from cross-contamination and safeguard against dangerous pollutants. The cultures' preservation in lyophilized or frozen form gives researchers the time they need to carry out thorough quality control checks, such as assessments for activity, identification, and purity. The regulatory abilities are crucial to guaranteeing adherence to the pertinent national and international laws and standards, both for the culture manufacturer and the food consumers. Starter culture producers differ in size from tiny businesses that offer solely cultures to food sectors that subscribe to them to major international businesses that supply other food ingredients. The startup culture sector experiences innovation driven by opportunity and demand. The biological sciences are advancing at a very fast rate, opening up new possibilities every day.

Our capacity to comprehend intricate biological systems has changed as a result of the development of techniques for gathering and analyzing massive volumes of data. The demand for new items serves as another factor fostering innovation. New techniques for crop preservation are desperately needed to prevent deterioration before consumption following harvest. We need improved ways to increase shelf life and prevent rotting in highly developed countries as well as in the less developed parts of the world. Some of the issues can be resolved by bioprotective microorganisms and food fermentation. The market would also welcome innovative goods in probiotic cultures with clear modes of action and specific health advantages. In the process of fermentation, microbes use the oxidation of carbohydrates and associated molecules to produce energy in the absence of oxygen. It has been a vital method for food preservation since the dawn of humanity. Fruits, vegetables, cereals, milk, and other raw materials have undergone unique processing to encourage the growth of advantageous microbes while preventing the formation of harmful and degrading microorganisms. Food will be preserved throughout fermentation, which will also improve the product's flavor, aroma, texture, and nutritional content. The production of lactic acid as well as acetic acids and, occasionally, antibacterial compounds, is what causes the preservation effect. It's crucial that the fermenting microbes use all of the fermentable carbohydrates before the unfavorable microorganisms can use/consume them. Product shelf life will be extended by the addition of salt by reducing water activity (Belz et al., 2012). The majority of the time, plant materials are the natural habitats of lactic acid bacteria, yeast, and molds. However, depending on the type of plant, the climate, and the nutrients present in the raw material, the type of organisms might vary greatly. Some fermentations require different kinds of microbes at different phases of the fermentation process.

2.9 PASTEURIZATION

When milk was pasteurized in the late 19th century, it was found that bacteria were required for the souring of milk for butter manufacturing. Bacterium lactis was the name of the pure culture that Joseph Lister isolated from sour milk in 1878. This bacterium was categorized by Orla-Jensen in 1919 as Streptococcus lactis; it is now categorized as Lactococcus lactis subsp. lactis. Vilhelm Storch in Denmark, Herman Weigmann in Germany, and H.W. Conn in the US independently introduced the addition of pure cultures to milk to enhance fermentation soon after this discovery (Behera et al., 2018).

Storch received a U.S. patent in 1896 for the formation of starting cultures. Because they primarily make lactic acid by catabolizing milk sugar (lactose), the majority of the bacteria used for milk fermentations were known as lactic acid bacteria (LAB), and because they were employed to start or initiate the fermentation, they were known as starters or starter cultures.

The idea of using starter cultures to make fermented foods, particularly bread, beer, fermented milk products, and cheeses, as well as sausages, has drawn more and more attention over time. The raw material treatment and spontaneous fermentation will promote the development of local flora. In the majority of spontaneous fermentations, a microbial succession occurs; frequently, LAB will predominate at first, followed by different species of yeast. Molds can only grow aerobically, which restricts the types of fermented foods in which they can be found. LAB creates lactic acid and other antimicrobial compounds that prevent the growth of harmful bacteria. They also reduce the amount of sugar in a product, extending its shelf life.

Alcohols and aroma components are the main products of yeast. Molds typically contribute to fermentation by creating both intracellular and external proteolytic and lipolytic enzymes, which have a substantial impact on the flavor and texture of the final product. Back-slopping is the practice of inoculating a new batch of a fermented product with a portion of a prior batch (Holzapfel, 2002).

This process promotes a quicker and more consistent fermentation than happens with spontaneous fermentation and generates a greater initial quantity of beneficial bacteria than is present in raw material. This process undoubtedly also encourages the development of bacteria that produce antimicrobial agents, resulting in the constant expansion of the same bacteria. Homemade fermentation of milk, vegetables, and cereals are examples of back-slopping. The addition of starter cultures is frequently used when it is possible to heat the raw material and inactivate the native flora, allowing the growth of starter microorganisms. However, sometimes heat-treating the raw material (such as fruits and vegetables) is not possible without changing the texture of the finished product. However, the presence of starter cultures, particularly those that include a bacteriocin-producing strain alone or in conjunction with particular bacteriocin-resistant bacteria, may increase the likelihood that the desired flora would predominate in plant fermentations. Single-strain starter cultures are mostly employed for the fermentation of yeasts, molds, and some dairy products, such as sausages and sauerkraut, as well as wine and beer. Wine, bread, sausages, and dairy products need different starter cultures. Mixed undefined bacterial starter cultures, also known as traditional starters, are mostly utilized in the dairy industry.

Lactic acid bacteria are considered to be the most significant bacterium for food fermentation. These bacteria are gram-positive rods or cocci, catalase-negative, microaerophilic, and typically nonmotile. From the fermentation of glucose, they mostly produce lactic acid. The taxonomy of bacteria has been impacted by the use of DNA technology and molecular typing techniques. Lactic acid is the primary end product of homofermentative LAB, but heterofermentative LAB also produces acetic acid, CO_2, and small amounts of lactate. Brevibacterium and Propionibacterium are utilized as secondary or adjunct cultures in food and beverage fermentations, along with other bacteria such as Acetobacter, Bifidobacterium, Micrococcus, and Staphylococcus.

As auxiliary cultures, some Lactobacillus species are also utilized. North and East Europe is the origin of mesophilic starting cultures. In particular, they are made up of Leuconostocmesenteroides subsp. cremoris, Leuconostoc lactis, Lactococcus lactis subsp. cremoris, Lactococcus lactis subsp. lactis biovar. diacetylactis, and Lactococcus lactis. Particularly, L. cremoris can quickly acidify milk. Citrate can be catabolized into CO_2 and diacetyl by L. diacetylactis, Leuconostocmesenteroides subsp. cremoris, and Leuconostoc lactis. The formation of holes/pores in cheeses is caused by CO_2, and many cheeses and fermented milk products depend on diacetyl, which gives butter its distinctive flavor (Curic et al., 1999). The organisms that create diacetyl are frequently referred to as aroma producers. The prevalent mesophilic lactobacilli present in various kinds of cheese and occasionally utilized as supplementary cultures are Lactobacillus paracasei (Figure 2.6) and Lactobacillus casei (Figure 2.7).

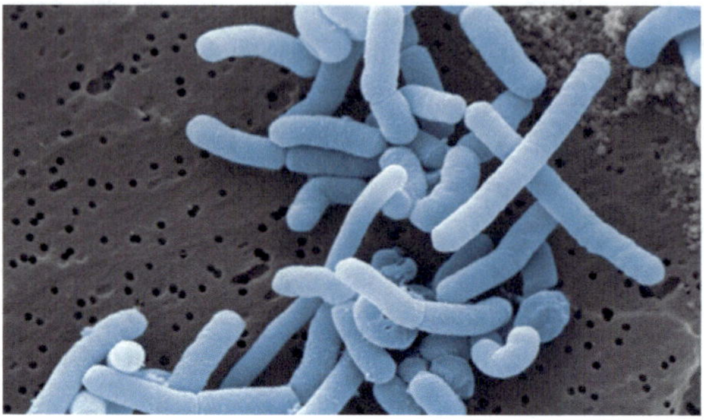

Figure 2.6. Lactobacillus paracasei.

*(**Source:** https://upload.wikimedia.org/wikipedia/commons/8/86/Lactobacillus_paracasei.jpg, Creative Commons License)*

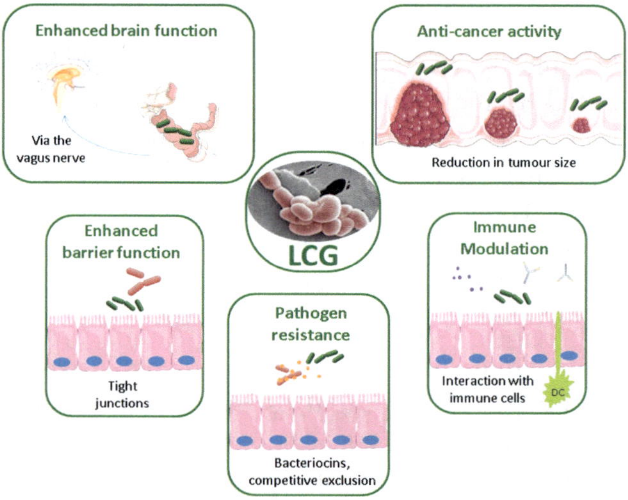

Figure 2.7. Lactobacillus casei.

(Source: Hill D, Sugrue I, Tobin C, Hill C, Stanton C and Ross RP (2018) The Lactobacillus casei Group: History and Health Related Applications. Front. Microbiol. 9:2107. doi: 10.3389/fmicb.2018.02107, Creative Commons License)

2.10 CHEESE

Many cheese types also contain Lactobacillusrhamnosus, Lactobacillus plantarum, and Lactobacillus curvatus. Some Swiss-style cheeses, including

Emmental, Gruyere, and Comte, use the dairy bacteria genera Propionibacterium named Propionibacterium shermanii and Propionibacterium freudenreichii to slowly catabolize lactate into propionate, acetate, and CO_2. This is crucial for the cheese flavor and hole/pore formation. In several kinds of cheese, Micrococcaceae and Brevibacterium are employed as surface flora; they are crucial for cheese ripening(Cahill, 1990). Mixed starting cultures of several types have been created. Cheddar cheese is the most prevalent cheese manufactured. A starter culture of Lactococcus lactis subsp. lactis and Lactococcus lactis subsp. cremoris, with or without Streptococcus thermophilus, is frequently used to make it. A mixed O-culture, TK_5, has also been created. The genesis of thermophilic cultures is in South and East Europe. Streptococcus thermophilus, Lactobacillus delbrueckii subsp. bulgaricus and Lactobacillus lactis and Lactobacillus helveticus make up the majority of them. The thermophilic LAB is employed as cheese additives or for quick acidification. Lactose is broken down into lactate and galactose, which are then released by Streptococcus thermophilus, Lactobacillus delbrueckii subsp. bulgaricus, and some Lactobacillus delbrueckii subsp. lactis. Galactose can be used by Lactobacillus helveticus as a carbon source, eliminating any remaining galactose. Some strains of Lactobacillus helveticus are highly proteolytic, which affects the flavor and consistency of cheese. The yogurt-specific flavoring acetaldehyde can also be produced by thermophilic LAB (Lubbers et al., 2004). These starter cultures come in different combinations. Mixed undefined cultures as well as specified single or multiple starting strain cultures are both employed. Although many cooked kinds of cheese are made with mixed cultures, mozzarella and yogurt are frequently prepared using single or multiple strain starters that contain one or more S. thermophilus and Lactobacillus delbrueckii subsp. bulgaricus strains. With or without rennet, they can be employed as milk cultures.

Other cheeses or milk products are produced using starting cultures that are mesophilic or thermophilic LAB or by adding additional LAB such as Enterococcus faecium, Lactobacillus acidophilus, or other species of bacteria (e.g. Bifidobacterium). Pure bacterial starter cultures have the drawback of being more susceptible to bacteriophages than mixed starters or native flora, where there are always strains that can withstand phage attacks and continue fermentation. The lactic acid fermentation of milk used to make cheese is particularly vulnerable to phage contamination. It is advised to utilize starting cultures resistant to phages, maintain a high degree of cleanliness and sanitation, use closed vials with filters, use equipment and facilities that have been properly designed and developed, restrict human access, and practice strict personal hygiene. Lactic acid, acetic acid, alcohol fermentations or a combination of these fermentation types are involved in plant fermentations. Lactobacilli and Pediococcus can also engage in the fermentation of alcohol, but yeast (Saccharomyces cerevisiae) and fungus (Aspergillus oryzae) are the predominant species. Vinegar is made via a two-step fermentation process called acetic acid fermentation, in which the first stage involves an alcoholic fermentation and in the second stage, there will be oxidation of ethanol via acetaldehyde to acetic acid (Yuan et al., 2013). Rice, potatoes, or grapes are the usual raw materials. For the manufacturing of vinegar, various subspecies of Acetobacter and Gluconobacteroxydans are utilized. In acetic acid fermentation, pure cultures are not frequently used. Yeast and mold fermentation is followed by lactic acid fermentation, which involves numerous LAB species that are active at various phases of the fermentation process in traditional fermentations of vegetables, fruits, and grains. The main microbes are Leuconostocmesenteroides and Lactobacillus plantarum, however many other LAB, like Pediococcus and other Lactobacillus species, may also be present.

2.11 VEGETABLE FERMENTATION

Vegetable fermentation is challenging to regulate because it is dependent on the quality of the starting material, the conditions of harvesting, and the temperature, all of which are essential for fostering the growth of the desired microorganisms (Marco et al., 2021). Vegetable fermentation frequently starts with the addition of salt, either in the form of dried salt i.e., brining, and the establishment of an anaerobic environment. One challenge is that raw vegetables are typically unable to be pasteurized without the resulting texture being negatively impacted. Another problem is that the fermentation of vegetables frequently involves a very intricate process in which numerous bacteria compete with one another in many ways. Commercial production of many distinct plant fermentations takes place mostly on a limited scale. Only olives, pickled cucumbers, sauerkraut, and kimchi are currently produced in significant quantities. Several strains of Lactobacillus were used for sauerkraut production, including Lactobacillus plantarum, Lactobacillus curvatus, and Leuconostoc mesentery alone or in combination with Lactococcus lactis. In addition to being used successfully for pickling cucumbers, Lactobacillus plantarum, Lactobacillus pentosus, and Pediococcuspentosaceus have been employed in the fermentation of olives. Different grains and legumes, such as soybeans, lupins, peas, and lentils, can be fermented. This fermentation does not require the addition of starter cultures; instead, it occurs naturally or by back-slopping. However, to make bread from wheat or rye, starter cultures are employed.

Yeasts are introduced into fermentation processes either through the raw ingredients or the process machinery. When yeasts are employed as starter cultures, they are typically employed as single cultures and can be added either at the beginning of the fermentation process or later on to ensure optimal odor generation. The majority of yeast species can flourish in both aerobic and anaerobic environments. Yeasts use carbohydrates as their main source of carbon, converting them into alcohols, CO_2, and a variety of secondary metabolites like esters, organic acids, aldehydes, and ketones. The majority of yeasts that are used in the fermentation of foods and drinks are ascomycetous yeasts. Saccharomyces cerevisiae is unquestionably the yeast species that has been most thoroughly described. Worldwide, this species is the most widely used yeast starter culture for the production of bread, wine, beer, cheese, and other products. Saccharomyces pastorianus, which is used to make lager beer, and Debaryomyceshansenii, which is used to make cheese and fermented meat, are two other important yeast species (Rainieri et al., 2006). Other types of yeast are potential starter cultures and do frequently appear in large numbers during spontaneous fermentations. The evolution of DNA technologies and molecular typing methods has had a considerable impact on the taxonomic position of many yeast genera in recent decades, and more reorganizations are anticipated in the future. There are several advantages to switching from spontaneous to controlled fermentation, therefore there seems to be an increase in interest in using yeasts as purified starting cultures to generate novel food items as well as to regulate already-existing fermentation processes. In controlled fermentation, it is important to take into account any probiotic properties or potential pathogenic hazards in addition to the habitat of the yeast species and its varied activities. Procedures for subspecies-level typing must be introduced, and the taxonomic status of the starter cultures must be specified.

2.12 STARTER CULTURES

The use of starter cultures of brewing yeast based on single cell cultures has a long history in the brewing industry. Up to 1,000 unique brewing yeast strains have been identified worldwide (Legras et al., 2007). The technological characteristics of the various brewing yeast strains include odor production, oxygen need, and reproduction. Maltose is the most prevalent carbon source throughout the fermentation process in brewing, however, sucrose, glucose, fructose, and maltotriose will be present and used. The capacity of different strains of brewing yeast to use maltose has been demonstrated, and genotypic variations in the number of maltose transporter genes have been recorded. Ale yeasts, also known as top-fermenting yeasts, and lager yeasts, sometimes known as bottom-fermenting yeasts, are both forms of Saccharomyces yeasts that contribute to the fermentation of beer. S. cerevisiae has been used to categorize ale yeasts, whereas Saccharomyces carlsbergensis, S. uvarum, and S. cerevisiae have been used to refer to lager yeasts. Although lager yeasts are still frequently referred to as Saccharomyces carlsbergensis, the development of molecular typing techniques has revealed several genetic differences between ale and lager brewing yeasts. Lager yeasts are now classified under Saccharomyces pastorianus group. Regarding the evolutionary relationship between lager yeasts and other yeast in the genus Saccharomyces, there is still considerable uncertainty. Lager yeasts have parts of two divergent genomes, one from S. cerevisiae and one from another Saccharomyces species, most likely S. bayanus, or a particular strain of Saccharomyces monacensis, which according to recent taxonomic keys, now also belongs to S. pastorianus.

In addition to its usage as an industrial starter culture, S. cerevisiae has been isolated from a range of different spontaneously fermented beers or beer-like liquids. Starter cultures of S. cerevisiae have also reportedly been employed in the production of South African sorghum beer (Sawadogo-Lingani et al., 2021). Wine is often made through spontaneous fermentation, which has been shown to involve several different yeast species. According to the grape variety, climate, soil quality, development, physical quality, and quantity of fungicides used in the vineyards, the prevalent microorganisms on the grapes differ. However, Kloeckera and Hanseniaspora are claimed to be the two most common yeast genera found on grapes, whereas Saccharomyces cerevisiae is either not present or is present only in very small amounts on healthy, undamaged berries. Candida, Cryptococcus, Debaryomyces, Dekkera, Hanseniaspora, Kluyveromyces, Metschnikowia, Pichia, Rhodotorula, Saccharomyces, Saccharomycodes, Schizosaccharomyces, and Zygosaccharomyces are some of the yeast genera connected to winemaking. The spontaneous wine fermentation process involves a natural yeast selection because of the low pH and high sugar content in grape juice. Yeast from the genera Candida, Hanseniaspora, and Kloeckera will predominate in the early stages of fermentation, followed by species of Metschnikowia and Pichia. S. cerevisiae strains that can tolerate alcohol will predominate in the later phases. The use of starter cultures in winemaking has seen an increase in attention over the past few decades, and the majority of large-scale processes are now carried out using starter cultures, mostly S. cerevisiae. For the manufacture of various types of wine, S. cerevisiae has been found to have many physiological variations, and S. bayanus has been employed as a starter culture, particularly for wine that is partially produced at low temperatures. Other genera besides Saccharomyces have recently been reported to be helpful in wine production to improve the taste and flavor of the wine.

Other than wine, yeasts are used to make a variety of unique alcoholic beverages, including a sizable number of native alcoholic beverages made via spontaneous fermentation. Several yeast species, including Pichia anomala and Pichia membranifaciens yeast, have been documented to be engaged in this secondary fermentation at its early stages. The yeast produces a surface film on top of the wine. However, a microbiological succession occurs during the fermentation, and S. cerevisiae is currently thought to be the most significant flour yeast. Non-Saccharomyces yeasts play a significant role in the early phases of native, spontaneously fermented beverages. However, S. cerevisiae will predominate in the later stages of the fermentations, as is the case in the brewing of sake, where isolates of S. cerevisiae, formerly known as Saccharomyces sake, are utilized. Baker's yeast has long been used in Europe and the US as a starter culture for the synthesis of a wide range of bread made using wheat. Back-slopping is a method of making bread that is common in Scandinavia and northern Europe. It appears that the European custom of making bread has now spread to every continent, including Southeast Asia and Africa. Hansen is always the predominant yeast utilized as a starter culture. Saccharomyces exiguus has been utilized as an alternative to baker's yeast.

Baker's yeast is made in large quantities. The total annual production is 2 million tonnes, and it has been stated that there has been a yearly growth of about 4 percent (Sudhakar et al., 2016). Using a fed-batch technique with aerobic conditions and low carbohydrate contents, baker's yeast is propagated. Optimization of baker's yeasts appears to concentrate on elements distinct from those of brewing yeasts, except the consumption of carbohydrates. Compared to brewing yeasts, it seems like there are fewer commercially available baker's yeasts. The crucial technological traits of baker's yeasts appear to be biomass yield, CO_2 production, influence on dough structure, aroma production, and high resistance to stress conditions like oxidative stress, drying, freezing, and thawing. These traits are in addition to the efficient utilization of the maltose, maltotriose, and other fermentable carbohydrates present in the dough. As a result, high yeast trehalose concentrations are required as a preventative measure. It is important to consider the adaptability to various environments, including those in the dough and those during manufacturing and propagation. In regions where typical European-style bread production is not practiced, such as a large portion of Africa, fermented grains other than classic wheat bread are common. Products made from fermented maize, millet, and sorghum are examples of native fermented grains. These compounds are typically produced via fermentation, with LAB coexisting alongside S. cerevisiae as the main yeast species.

2.13 YEAST

Yeasts are employed in the manufacturing of cheese for dairy goods, though they can also be used to make fermented milk. Although many products are still made by back-slopping or fermentation, yeasts are still frequently used as single-starter cultures. In many dairy products, the starter culture that was employed and indigenous flora are mixed. Most often, yeasts are utilized as supplementary starting cultures to boost the production of aromas or to promote the growth of other microorganisms. The roles of yeasts during the manufacturing of cheese and their impact on the cheese quality are generally understudied. For the production of surface-ripened cheeses such as Brick, Limburger, Port Salut, Taleggio, Tilsitter, Trappist, and Danish Danbo cheese, the incidence and functions of yeasts have been specifically investigated. It is discovered that these cheeses have a mixed flora on their surface that includes both yeasts and bacteria. Debaryomyceshansenii, an osmotolerant yeast, has been discovered almost exclusively in Danish Danbo cheese, although other surface-ripened cheeses have also been reported to contain Candida zeylanoides, Yarrowialipolytica, Kluyveromyces lactis, (Figure 2.8) and other yeasts (Breuer & Harms, 2006).

By degrading lactate, the yeasts start the ripening process. This raises the pH of the cheese surface and encourages the establishment of acid-sensitive bacterial flora, including Brevibacterium linens. It should be noted that the species D. hansenii has two varieties: D. hansenii var. hansenii and D. hansenii var. fabryki. It appears that D. hansenii var. hansenii is the common variant encountered in cheese production, and isolates of this variety have been introduced as starting culture.

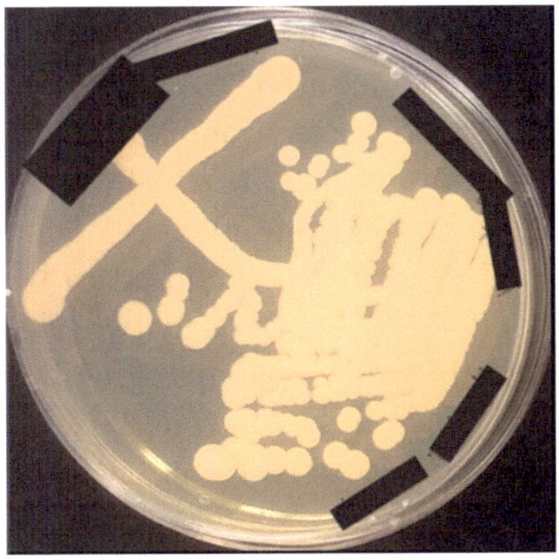

Figure 2.8. Kluyveromyces lactis.

*(**Source:** https://upload.wikimedia.org/wikipedia/commons/e/e8/Kluyveromyces_lactis_plate.png, Creative Commons License)*

It has been discovered that K. lactis is widely distributed in soft cheeses like Camembert. S. cerevisiae has been employed as a starter culture for the synthesis of Gorgonzola, but it also appears to naturally occur, along with other yeasts, as an essential component of the microflora of both blue-veined cheeses and some varieties of soft cheeses. Saccharomyces cerevisiae influences the production of aromas, has some proteolytic activity, and encourages the mycelial proliferation and conidia formation of Penicillium roqueforti.

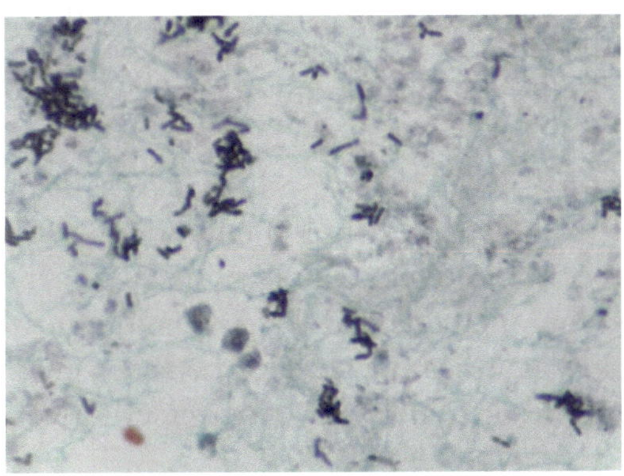

Figure 2.9. Galactomycesgeotrichum.

*(**Source:** https://upload.wikimedia.org/wikipedia/commons/d/d3/Geotrichosis_Case_151%282%29.jpg, Creative Commons License)*

Galactomycesgeotrichum (Geotrichumcandidum), (Figure 2.9) once assumed to be a mold, is now thought to be a yeast species. G. geotrichum is connected to dairy products and the environment. It is recognized as a starter culture for various varieties of surface-ripened cheeses, including cheeses like the French St. Albray, as well as mold-ripened cheeses. G. geotrichum, on the other hand, has also been demonstrated to be a potential spoiling organism. For the production of surface-ripened cheeses, G. geotrichum is combined with bacteria like Penicillium camemberti or Brevibacterium linens. Yarrowialipolytica is frequently found in soft, blue-veined, and surface-ripened cheeses despite not being utilized as a starting culture in the dairy industry. When used as a starter culture for making cheese, Y. lipolytica has a significant lipolytic and proteolytic activity that could be challenging to manage. It is known that Y. lipolytica produces brownish color in cheese. Yeasts from the genera Candida, Galactomyces, Kluyveromyces, Saccharomyces, and Torulaspora are used to produce fermented milk. Viili, a fermented milk product from Scandinavia, is made using G. geotrichum (G. candidum) as a commercial starter culture and milky kefir is made using Saccharomyces unisporus and Kluyveromycesmarxianus (Candida kefyr) as commercial starter cultures. In Europe, starter cultures of mold are largely utilized to produce cheese, and spontaneous mold fermentations are employed globally to produce a range of native foods (Sunesen & Stahnke, 2003). Molds can only be used in environments with oxygen, which restricts their uses. Molds are generally recognized as efficient enzyme makers, and many of them exhibit significant proteolytic and lipolytic capabilities as well as extracellular glucoamylase activity. It is known that molds produce a wide range of distinct compounds, the most well-known of which are alcohol and organic acids. Molds can develop in a protein-rich environment without carbohydrates, where they utilize amino acids as a carbon source, although they prefer to grow on carbohydrates. Molds are classified according to their micromorphology and how they develop on various surfaces, but they can also be classified according to the secondary metabolites they produce. The use of molecular typing methods and DNA technologies for molds has not progressed as quickly as it has for yeasts, and the typing of molds used as starter cultures still relies heavily on phenotypic criteria.

2.14 PENICILLIUM

Penicillium roqueforti and Penicillium camemberti are two molds that have a long history of usage in the manufacture of cheese (Ropars et al., 2020).

They were first used as starter cultures at the beginning of the 19th century. P. roqueforti is employed as a supplemental starter culture in the production of blue-veined cheeses, but it can also appear accidentally in a variety of other foods and is regarded as a contaminant for some kinds of cheese. P. roqueforti is suited for the manufacture of blue-veined cheese due to its ability to thrive at high NaCl, low O_2, and moderately high CO_2 concentrations. P. roqueforti produces a variety of extracellular peptidases and proteinases during cheese maturation. The proteolytic and lipolytic activity of various commercial strains of P. roqueforti has been found to differ significantly. In particular, the lipolytic enzymes impart the distinctive flavor and taste of blue-veined cheeses due to their production of methyl ketones. P. camemberti is a starting culture used to make Camembert and other similar types of surface-ripened cheeses. It is primarily utilized in the manufacturing of cheese and is hardly ever found in other foods or the environment. Due to its high NaCl tolerance, it can grow on the surface of the cheese. It breaks down lactate on the cheese's surface, raising the pH. Following the breakdown of casein by P. camemberti proteinases, which occurs when lactate is no longer present, the pH rises once more and ammonia is released. P. camemberti has strong proteolytic activity in addition to producing lipases that are essential for aroma production.

It is mostly concentrated in European countries like Italy, Spain, France, Germany, and Hungary where molds are used as starter cultures for the fermentation of meat. The fermentation is mostly caused by Penicillium spp., and the products produced are typically sausages and ham. Penicillium nalgiovense, Penicillium chrysogenum, and Penicillium camemberti are often used molds for meat fermentation, however, other species may also be employed. Penicillium aurantiogriseum has been linked to the spontaneous fermentation of dry sausages, Penicillium commune, Penicillium chrysogenum, Penicillium echinulatum, and Penicillium expansum have been found during the spontaneous fermentation of dry-cured Iberian ham. Penicillium commune and Penicillium olsonii have also been linked to the spontaneous fermentation of Spanish meat sausage. It was discovered that the two prevalent species used in the synthesis of traditional Tyrolean smoked and cured ham were Eurotium rubrum and Penicillium solitum. Although Aspergillus spp. are not employed as starter cultures for the fermentation of meat, they may be seen during the manufacture of cured ham because they may grow in environments with less water. Molds not only affect the product's appearance but also its distinctive flavor and scent by producing extracellular proteinases and lipases. Molds also have an antioxidant effect and can stop the growth of undesirable microbes. P. chrysogenum and P. nalgiovense strains are both recognized penicillin producers, and at least the latter has been demonstrated to be capable of producing penicillin when growing on meat surfaces and secreting it into the final product (Houbraken et al., 2011). Therefore, it is crucial to carefully examine mold starter cultures using both chemical and biological testing to make sure that they do not produce mycotoxins or penicillin in the final product.

Although Botrytis cinerea development may be necessary for the manufacture of some types of wine, molds are not employed as starter cultures for winemaking. B. cinerea is generally undesirable since it will cause the grapes to rot. However, the development of B. cinerea on mature grapes is necessary for the production of sweet white wine, such as French Sauternes and Hungarian Tokay.

This causes overripening and dehydration, which increases the sugar content. The development of B. cinerea also gives the wine made from these grapes a distinctive flavor.

The yeast Pichia membranifaciens has been used to biologically regulate the growth of B. cinerea on mature grapes; its antagonistic impact against B. cinerea appears to be connected to the secretion of exo- and endo-h-1,3-glucanases (Masih & Paul, 2002). Foods that have undergone fermentation have long been accessible to various cultures. The acceptance of fermented foods differs culturally as well: A product that is highly appreciated by consumers in one country could not be in another. Numerous food items are fermented. Due to factors like food categories, product form and qualities, type of ingredients utilized, and cultural variation, manufacturing techniques for fermented goods vary greatly.

2.15 DAIRY PRODUCTS

Fermented dairy products are frequently made by nomads and nations that produce milk. Through cultural exchange, these products, which are very acceptable in these civilizations, are eventually embraced by other cultures.

It is widely believed that nomadic peoples were responsible for the discovery and early development of the majority of fermented dairy products. Today, the process of making a fermented dairy product can be extremely complex. The milk, microbes, and other components employed in the production process have an impact on the quality of a fermented dairy product. Milk composition is influenced by a variety of factors. Breed, feed, season, geography, and herb health are the variables that have an impact on the processing of milk products. To meet national standards, milk composition is standardized in industrialized nations. It is acknowledged, though, that standards in different nations may not be the same, and as a result, the composition of a given product may change. There are now accessible international agreements to standardize some products. However, due to the microorganisms and culturing techniques employed in their manufacture, goods made in various regions can still differ from one another.

CHAPTER

3 Microbiology of Fermented Foods

Contents

3.1 INTRODUCTION

Commercial microbial strains developed by precise genetic engineering protocols help in the production of fermentation products. For a long time, barley and grapes were used in the fermentation process to produce alcoholic beverages. The microbial strain and environmental factors are used in the production of by-products and desired products (Canete-Rodriguez et al., 2016). Enzymes have been utilized in the industry for a long time due to their catalytic activities, and their actions rely on the substrate, pH, inhibitors, and temperature. Every procedure needs to be improved. Enzymes have a wide range of uses in industries. Enzymes are used in a process that creates a lot of commercial goods. Enzymes have been employed to make cheese, yogurt, beer, and wine for thousands of years. Alcohol and carbon dioxide gas are produced as a result of yeast enzymes breaking down sugar. Biotechnology technique is mostly utilized in fermenter setup, control, product improvement, and characterizing chemical engineering processes by microorganisms and their output. Fermentation is a crucial component of biotechnology and is expanding quickly. Fermentation technology is extensively used in the pharmaceutical, food, and alcoholic beverage sectors on a large scale. Productivity is influenced by fermentation mechanisms, including strain types, media, and growth conditions.

When producers are aware of the biochemical changes occurring in fermented meals, they can expand the production of fermented goods. By manipulating the conditions, the production characteristics can change. By choosing modes of fermentation i.e., continuous, batch, and fed-batch, high productivity can be attained. Higher productivity can be gained from it because cell immobilization raises the concentration of biomass in the bioreactor, which also increases the concentration of the biocatalyst. The fermentation product recovery method needs to be assessed and optimized, and its economic constraint factor also needs to be improved. Examples of recovery techniques include extraction, homogenization, and filtration. Another phase in this process that is expensive is the purification of the finished product. Microbial end products might also include extracellular compounds, biomass, or intracellular products. Each process has its enzyme activity optimized, which depends on factors like pH, temperature, inhibitors, substrate, etc. Enzyme immobilization is used to save process costs because enzyme recovery is challenging (Christopher et al, 2014). Plants, mammals, and microorganisms can be used to isolate enzymes. However, microbial enzymes are chosen because of their accessibility and specificity. A variety of enzymes have been employed in pharmaceuticals, food, detergents, paper, and cosmetics. Proteases, lipases, pectin enzymes, milk clotting enzymes, and amylases are examples of commercially available enzymes.

3.2 INOCULATION

With the inoculation of a substrate, the environmental conditions must match those of the targeted microbe to produce the required yield. To achieve the desired biochemical reaction, mono-septic fermentations require a certain microbial variety. Other microorganisms are killed during sterilization, and a sterilized substrate is then employed. It is used to produce pharmaceutical goods like insulin (Figure 3.1).

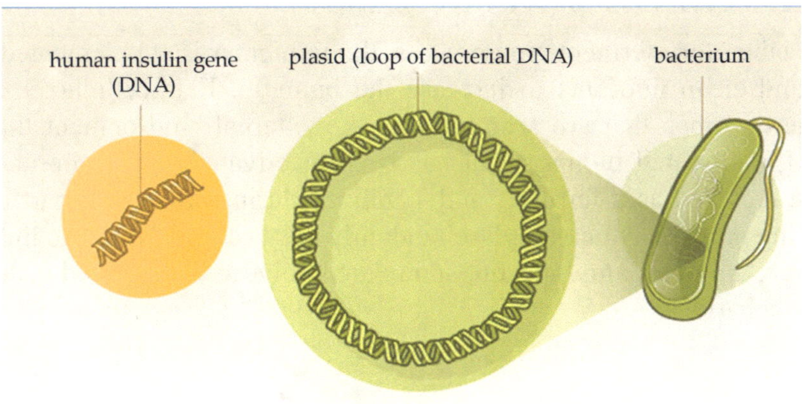

Figure 3.1. Insulin.

(**Source:** *https://www.nlm.nih.gov/exhibition/fromdnatobeer/exhibition-interactive/recombinant-DNA/alternative/recombinant-dna-technology-02.jpg, Creative Commons License)*

There are various microbial species involved in the treatment of biological waste and several food fermentations. Mixed-culture involvement is necessary. It is based on the components of microbial compatibility. This method of generating pharmaceutical proteins uses transgenic microorganisms or mammalian cell culture systems to offer affordable platforms for the manufacture of therapeutic proteins. These include drugs, blood proteins, antibodies, and vaccinations. NADH enables the cell to continuously produce energy via glycolysis by substrate-level phosphorylation and is recycled to NAD^+. To maintain energy balance, microbes are useful in producing a variety of end products or by-products. The food business used enzyme technology to purify fruit juice for the first time in the 1930s (Alvarez Garcia, 2018). Because microbial enzymes have so many uses, the food industry has been utilizing them for a long time. The leather industry also makes use of these enzymes. A fermenter, a sizable container with a volume of up to 1,000 cubic meters, is where fermentation takes place. Nutrients based on renewable raw materials like maize, sugars, and soya are the fermentation media, which should be sterilized. In the fermentation medium, the bulk of industrial enzymes is secreted by microorganisms to manage the breakdown of the carbon and nitrogen sources.

3.3 FERMENTATION PROCESS

In the laboratory fermentation process, the fermenter is supplemented with pure and clean nutrients to increase the biomass. Extracellular enzymes are the enzymes that are secreted into the external environment through cells. Lime-treated biomass that has been inactivated by microorganisms can be recycled as a fertilizer and stabilized during storage as part of the recycling process. Alpha-amylase randomly cleaves 1,4 glycosidic linkages of starch, amylose, amylopectin, and related polysaccharides and generates oligosaccharides.

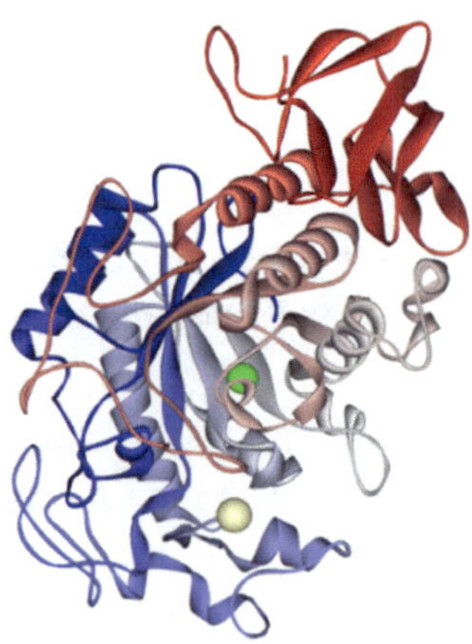

Figure 3.2. Amylases.

*(***Source:*** https://upload.wikimedia.org/wikipedia/commons/6/66/Salivary_alpha-amylase_1SMD.png, Creative Commons License)*

Amylases (Figure 3.2) are widely found in microbes, plants, and animals. To produce amylase from bacteria, the traditional method of submerged fermentation is frequently used. However, solid-state fermentation is employed to manufacture extracellular fungal amylase from low-cost agricultural products. Microbial amylolytic enzymes replace chemical starch hydrolysis in industrial sectors; thus, their demand is rising steadily (Olympia et al., 1995). Amylase has about 25% market share in the global enzyme industry. The market for industrial enzymes has increased dramatically over the years, and this growth pattern is expected to continue. There has been an increase in the manufacture of economically viable enzymes. Commercial enzyme production is carried out by many top multinational corporations. Amylase has a long history in the field of enzyme chemistry as a biocatalyst. The discovery of a starch-degrading enzyme in wheat by Russian chemist Kirchhoff in 1811 marked the beginning of the history of amylases. Anselme Payen and Jean-François Persoz, two French chemists, discovered an enzyme complex from barley in 1833 that they named diastase and could convert gelatinized starch into sugars, chiefly maltose. In 1894, Japanese researcher Jokichi Takamine succeeded in isolating fungal α-amylase from an Aspergillus oryzae surface culture on wheat bran (Bhanja Dey & Banerjee, 2015).

3.4 CARBOHYDRATES

All living things include carbohydrates, and each enzyme that splits them up is a member of a different family of glycosyl hydrolases (GH family). There are 18 classes (A-R) and 156 glycosyl hydrolase families (GH_1-GH_{156}). The GH families 3, 13, 14, 15, 31, 57, 119, and 126 contain starch-degrading enzymes, according to the carbohydrate-active enzymes database (CAZy). With more than 30 distinct enzymes, the α-amylase (O-glycosyl hydrolases) family is the largest family. The glycosyl hydrolases family 13 (GH_{13} family) is where the major enzyme, α-amylase, originates from (Lopez-Casado et al., 2008). For green technology, many biotech companies develop several beneficial enzymes. Extracellular enzymes are mostly produced by microbes, especially α-amylase. It is desirable to produce amylase utilizing microbes because they have a large production capacity and are less expensive. Both submerged fermentation (SmF) and solid-state fermentation (SSF) have been used to produce α-amylase. For optimum microbial growth and enzyme production, some physicochemical factors are being taken into consideration, including nutrient replenishment, medium pH, degree of aeration, and temperature regulation. SmF is often carried out in a liquid medium where microorganisms can proliferate in the presence of water. It is more appropriate for the production of α-amylase by bacterial species, primarily those belonging to the genus Bacillus. For this reason, synthetic medium with nitrogen sources, soluble starch, and other ingredients are employed, which raises the cost of production. However, it has certain advantages, such as ease of handling and greater control over process optimization parameters like temperature, pH, aeration, etc. A growing number of industries and researchers are now interested in SSF for the manufacture of enzymes due to its technical and economic benefits. This method is practical because the microbial fermentation was completed in an environment with limited water. SSF is the best option for producing extracellular enzymes from fungi on an industrial scale, particularly from filamentous fungi and yeast since they can flourish in an environment with minimal water activity. By creating turgor pressure at the tip of the mycelium, filamentous fungi can enter solid agro-industrial substrates.

3.5 PROCESS OPTIMIZATION

Process optimization takes into account several variables, including surface area, porosity, moisture content, particle size, and nutrient supplementation (Rao & Basak, 2021).

Due to the use of moist agricultural polymeric substrates for this purpose, such as wheat bran, rice bran, rice husk, cassava, sunflower meal, cottonseed meal, soybean meal, pearl millet, etc., SSF is the most advantageous approach in terms of production costs. Since agro-wastes are primarily destroyed by fire, applying them increases the likelihood of waste recycling and reduces pollution.

Despite these, SSF has many additional benefits, including its ease of use, low capital investment, high production rate, low levels of end product inhibition and catabolite repression, low wastewater output, better product recovery, low purification costs, and high-quality product extraction. By changing one variable at a time (OVAT), the traditional approach to process optimization has been carried out, in which several independent components and their levels are separately optimized (Kennedy & Krouse, 1999).

Response surface methodology (RSM) has recently been used to streamline the process of optimizing the culture medium for increased enzyme production in industrial fermenters. This procedure aids in locating the key influencing elements that boost the yield of α-amylase in reactors. Design of experiments (DOE) is used to operate RSM. The identification of the points at which the solution should be examined is the goal of DOE. This method is employed in microbiological research for process optimization, which may evaluate the specific effects of each process condition on overall effectiveness. RSM thus offers details regarding the optimal values of each variable, their interactions, and their influence on the product yield.

3.6 ARTIFICIAL NEURAL NETWORKS

Artificial neural networks (ANNs) are the current method for process optimization. An ANN comprises three different layers of connectivity (neurons). Inputs from the independent variables (neurons) make up the first layer; hidden neurons in the second and third layers stand in for nonlinear activation functions and output neurons, respectively (Desai et al., 2008). Each of the first layer's neurons is linked to a layer or layers of hidden neurons, which are linked to a level of output neurons at the bottom. The number of neurons needed in the hidden layer can be calculated using a trial-and-error method to reduce the deviation of predictions from experimental data.

The model can only be developed by ANN with a minimum of ten neurons. The algorithms were created based on the complicated relationships between input neurons and the observed outcome. It is feasible to create a multilayer perceptron (MLP) using computer software (MATLAB). The most important aspect of downstream processing is enzyme purification. After fermentation, purification procedures begin; their effectiveness is highly influenced by the technology available, consumer demand, and financial investment.

By using chromatographic methods, the majority of extracellular enzymes are purified until they are homogeneous. The traditional multi-step purification method is time-consuming, costly, and always runs the risk of not producing the intended product with a low yield. After the advancement of purification systems with quick, economical, and efficient protocols with fewer processing stages, large-scale, cost-effective bulk enzyme purification for commercial applications was established. The most efficient methods for the industrial-scale purification of amylase are forced affinity chromatography, expanded bed/fluidized bed chromatography, high-speed counter-current chromatography (HSCCC), and magnetic affinity adsorption chromatography (Figure 3.3).

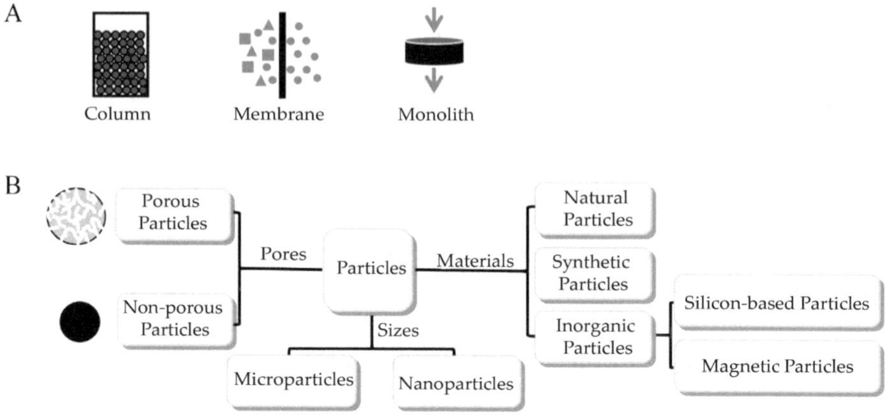

Figure 3.3. Chromatographic matrix for use in antibody affinity purification. (A) Formats of the stationary phase; (B) Classification of particles based on different criteria.

*(**Source:** Xue A, Fan S. Matrices and Affinity Ligands for Antibody Purification and Corresponding Applications in Radiotherapy. Biomolecules. 2022; 12(6):821. https://doi.org/10.3390/biom12060821, Creative Commons License)*

With changes in pH, the enzymatic activity fluctuates. While an excessive pH will always disrupt the protein structure, a specific pH stabilizes the enzyme's tertiary structure. The primary determinant of catalytic activity is the ionization of the side chains of certain amino acids. Thus, the stability of the enzyme, substrate binding, and catalytic activity depends on the pH of the reaction mixture being controlled. The pKa of the proton acceptor and proton donor determines the ideal pH. The pH profile spectrum for maximum activity is determined by the range between the pKa of the proton acceptor and proton donor. Every enzyme has an optimal temperature where it operates at its highest level. The range of temperatures that promote α-amylase activity is 25 to 100°C. The kinetic energy of collisions between reaction groups increases as temperature rises in an enzyme-catalyzed process. Heat causes the protein to become denatured at high temperatures. Understanding protein stability, folding, and the development of more stable enzymes for commercial usage will be aided by research into the thermostability of α-amylase. Thermostable α-amylases can continue to work at temperatures as high as 80°C to 90°C and 100°C to 110°C to liquefy and gelatinize starch, respectively (Maeda et al., 2003). The number of hydrogen bonds, ionic interactions, hydrophobic interactions, shorter surface loops, disulfide bonds, metal ion binding, rigid and compact packing, GC-rich codons, and the ratio of charged to uncharged amino acids are a few characteristics that influence thermostability at the molecular level. In addition to these, the presence of stabilizers such as calcium ions, substrate, and others affects thermostability. The thermostability of amylase is typically linked to Ca^{++}, and there may be one or two inherent Ca^{2+} ions near the active site. Mutational analysis showed that the Bacillus amyloliquefaciens α-amylase has the calcium-binding residues Asp_{231}, Asp_{233}, and Asp_{438}.

3.7 PROTEINS

The alpha helix of a thermostable protein has more beta strands than pro, his, or cysteine, and prefers arginine because it has available ionic charges. Glu, Arg, Tyr, Asp, and Lys have a larger surface area than Ala, Asn, Gln, Thr, Ser, and Val, which increases ionic interactions, structural stability, and thermostability. Ala, Gly, Trp, and Tyr are scattered throughout the protein region. Some of these are the result of codons with GC. A thermostable protein, which also contains more Val, Ile, and Leu, must have β-strands to function. These non-polar amino acids play an important role in cavity-filling due to their aliphatic side chains, which are also closely linked to hydrophobic interaction.

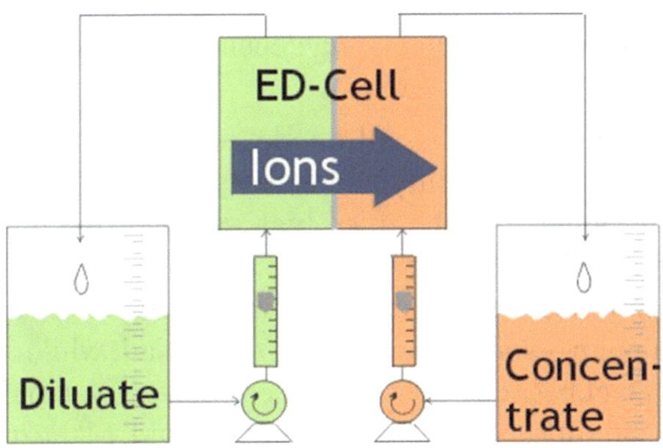

Figure 3.4. Electrodialysis.

*(**Source:** CC BY-SA 3.0, https://en.wikipedia.org/w/index.php?curid=3721420, Creative Commons License)*

The presence of ionic bonds between subunits indicates the protein stability and thermostability. Any protein must be flexible in its conformation to function, yet rigidity is necessary to preserve a protein's three-dimensional (3D) structure. The flexibility of proteins is significantly influenced by temperature. The natural structure of the protein is stabilized by some non-covalent interactions. By reducing the strength of the non-covalent connections, increased temperature triggers the unfolding of proteins. These effects improve conformational flexibility while reducing rigidity. At higher temperatures, α-amylase exhibits reversible unfolding and returns to its original structure upon cooling. α-Amylase is a metalloenzyme that contains calcium ions.

Electrodialysis (Figure 3.4) or the chelating agent EDTA (ethylene-diamine-tetra-acetate) can be used to remove calcium ions. Calcium-free enzymes lose their activity, but Ca^{2+} ions can make them active again. Divalent cations including Sr^{2+}, Mg^{2+}, and Ba^{2+} can also help to revive the activity of α-amylase that has been inactivated by EDTA. Calcium interacts with negatively charged residues like aspartic and glutamic acid to stabilize the structure of α-amylases. Calcium can bind to some secondary sites in amylase. These are also found close to catalytic residues (Asp, Glu), which are linked to catalytic efficiency. The melting temperature (T_m), mid-transition concentration (C_m), and overall stability of amylase are increased by calcium binding. Numerous studies have been presented in favor of Ca^{++} ion-independent thermostable amylase as an alternative to calcium dependence. There are Ca^{2+} inhibited amylases as well as Ca^{2+} dependent or independent amylases. Despite being dependent on calcium, sodium ions can also retain α-amylase in nature.

At the junction of domains, all amylases contain a single conserved calcium ion. The enzyme stability and activity depend on this calcium ion. A sodium ion is included in the microbial amylase structural component. One sodium ion and three to four Ca^{2+} ions are carried by the α-amylases of Bacillus species. These metal ions construct a bridge between domains in the form of a calcium—sodium—calcium metal triad, while a third calcium ion mediates contact between domains. Thermal inactivation is avoided and conformational stability is provided by the metal triad. Many α-amylases also have a chloride ion in the active site, which raises the pKa of the hydrogen-donating residue in the active site and improves catalytic efficiency. Chloride ion binding increases the conserved calcium ion affinity, which ultimately causes conformational changes at the active site. The existence of four highly conserved areas in the internal structure of the α-amylase family is an interesting characteristic. The development of the active center and the substrate-binding site depends on these regions. Histidine residue in the first region interacts with the substrate glucose residue; nucleophilic Asp is present in the second region; proton-donor Glu residue has been found in the third region; and histidine residue and asp residue may form hydrogen bonds with the substrate glucose residue in the fourth region. Acidic amino acid residues Asp, Glu, and Asp are located in the center of the TIM, where the catalytic site is located (Tamura et al., 1989). Comparative analysis of the nucleotide sequences of several microbial amylases has revealed that bacterial amylase contains more base pairs than TAKA amylase from Aspergillus oryzae, but the quantity of amino acid residues differs between the different organisms. In contrast, although their position number varies, the three amino acids that make up the catalytic triad are always present in the conserved areas. The catalytic residues Asp_{206}, Glu_{230}, and Asp_{297} in the TAKA amylase of Aspergillus oryzae are identical to Asp_{231}, Glu_{261} and Asp_{328} in the amylase of Bacillus licheniformis (BLA).

The C-terminal end of the β-strands in the TIM-barrel is where the active site cleft is found at the interface between domains. Starch may fit four to ten glucose units in the substrate binding cleft. Four highly conserved areas are used by amylase for substrate binding and catalysis. The substrate binding subsite is composed of residues that bind to the substrate. The catalytic activity is carried out by two aspartic acids and one glutamic acid. When the enzyme-substrate complex is formed, aspartic acid serves as a nucleophile and glutamic acid as an acid/base catalyst. To enhance the functional properties of industrially significant proteins (enzymes), protein engineering is frequently used. By carefully altering the amino acid sequence of protein-encoded genes depending on the mutation, the idea of protein engineering is successfully realized. One effective method for protein engineering that modifies a DNA sequence specifically is site-directed mutagenesis. With the help of this method, it is possible to isolate the desired mutants from millions or thousands of progeny while reducing the risks associated with crude mutagenesis in cells or organisms.

It was necessary to modify the amino acids in amylase to improve its specific activity, kinetic parameters like substrate affinity, catalytic efficiency, pH profile, thermostability, oxidation resistance capacity against oxidizing agents like iodate, heavy metals, sodium hypochlorite, activity against chelators like EDTA, EGTA, DTPA (diethylene triaminepentaacetic acid), zeolites, etc. The modified/engineered enzymes can be used in the food, detergent, textile, and paper industries. Several methods have been used to produce enzymes of higher grades. It is advantageous to substitute other amino acids for some oxidation-prone amino acids like methionine, tryptophan, cysteine, and histidine to increase the capacity for oxidation resistance. Leucine can improve the oxidative stabilities of Bacillus amylase.

By replacing Met_{197} with Ala in the wild-type amylase of Geobacillus stearothermophilus, an engineered form of amylase was also produced, which demonstrated significantly high resistance to chelator agents and chemical oxidation (Khemakhem et al., 2009). The likelihood of its usage in the detergent business rises with the improvement of these characteristics. The majority of chelator substances interact with sulfur compounds to produce inactive proteins. But among the non-polar amino acids, alanine has the lowest hydrophobicity and helps maintain the stability of protein structure. Disulfide bonds are added to the enzyme, which increases protein stability. The bond energy of disulfide bonds is higher than that of non-covalent bonds. Disulfide bonds naturally form between two cysteine molecules, and the portion of the protein chain makes it possible for cross-linking, which increases stability. However, due to dihedral angle restrictions, the production of cysteine disulfide bonds necessitates a considerable distance between the two cysteine molecules. The inclusion of non-canonical amino acids with long side chains thiol groups can occasionally produce a favorable environment for the formation of disulfide bonds as well as protein stability. As amylase is present in a folded or partially unfolded state under diverse conditions, especially at high temperatures, the introduction of proline in loop areas can stabilize the proteins by lowering the entropy level of the unfolded state. Proline residues must fit into the protein structure to be inserted, and arginine is typically chosen as the replacement for proline. The stability of an alkalophilic Bacillus amylase is increased by substituting proline for Arg_{124}.

3.8 ENZYMES

It is crucial to produce hybrid enzymes employing distinct homologous amylase genes from various bacteria to boost their thermostability and catalytic effectiveness. Bacillus licheniformis (BLA) and Bacillus amyloliquefaciens (BAA) have more than 70% identity in their α-amylase, according to a sequence alignment analysis (Khajeh et al., 2006). For the formation of the hybrid enzyme, these two species were utilized. Two areas of BLA were found to have increased thermostability when compared to BAA. Later, three more changes were made to BAA by deleting Arg_{176} and Gly_{177} and replacing Lys_{269} with Ala. These alterations increased BAA thermostability further. Other α-amylases of different Bacillus species showed similar effects on the enhancement of thermostability.

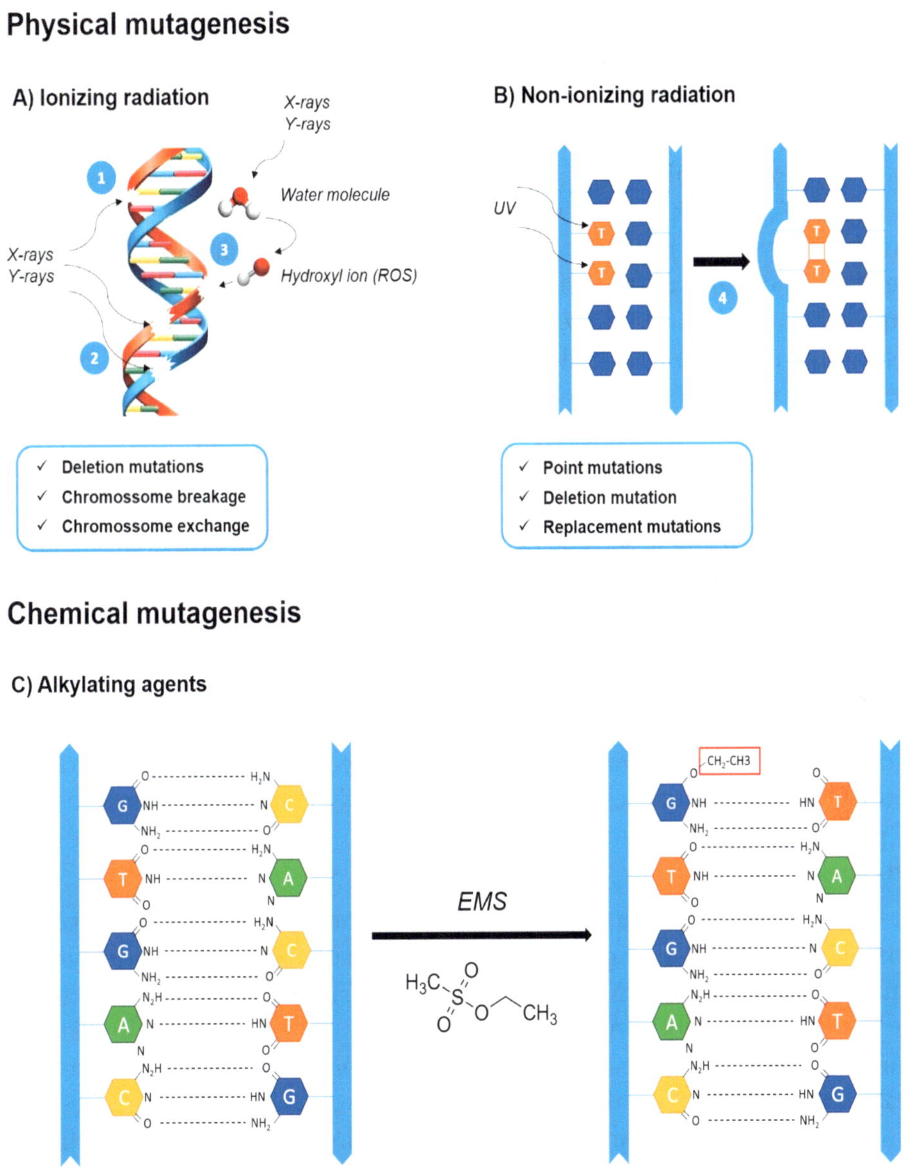

Figure 3.5. Random mutagenesis.

*(**Source:** Trovão M, Schüler LM, Machado A, Bombo G, Navalho S, Barros A, Pereira H, Silva J, Freitas F, Varela J. Random Mutagenesis as a Promising Tool for Microalgal Strain Improvement towards Industrial Production. Marine Drugs. 2022; 20(7):440. https://doi.org/10.3390/md20070440, Creative Commons License)*

Random mutagenesis (Figure 3.5) and high-throughput screening with error-prone PCR are used to design amylase. At various pH levels, α-amylase is used in many industrial industries. Many methods can be used to enhance pH profiles for optimum activity. The pH profile can be drastically altered by changing the amino acid Asn_{329} in Bacillus stearothermophilus amylase to Lys or Asp. It is known that asparagine is a deaminating amino acid, which weakens the structure of proteins. Lys or Asp substitution improves the likelihood that a salt bridge will form and interact with the solvent, improving enzyme stability. In the protein structure, helix capping is predominantly accomplished through the formation of hydrogen bonds at the ends of the helices. The N-cap is more favorable for negatively charged amino acids because it can balance the partial dipolar charge of the unpaired amide protons. The C-cap contains glycine, which can appear as the -L shape for stability. During protein folding, a cavity is mostly produced at the strand found at the turn and loop. The protein hydrophobic core grows as a result of this activity, which also stabilizes energy. The optimum product is obtained by replacing alanine and valine with non-polar, long aliphatic hydrocarbon chains containing amino acids like leucine and isoleucine. For the synthesis of commercial enzymes, pharmacological agents, food additives, etc., biotechnological methods have been used. The manufacturing of ethanol and biofuels, the manufacture of textiles, paper, detergents, the food and pharmaceutical industries, the preparation of animal feed, the recycling of paper, and the bioconversion of wastes are only a few applications for amylases.

The first stage in the manufacturing of branched dextrin, high fructose corn syrup (HFCS), oligosaccharide combination, and glucose syrup is liquefaction (Guzman - Maldonado et al., 1995). The unbalanced mixture of amylose and amylopectin known as starch creates a suspension in water. To start the liquefaction process, which results in the production of short-chain dextrin and a decrease in the viscosity of the starch suspension, commercially available thermostable amylases, primarily from Bacillus amyloliquefaciens and Bacillus licheniformis, are widely used in starch processing technology. Using the enzymes α-amylase, isoamylase, pullulanase, glucoamylase, β-glucosidase, and maltotetraose-producing amylase, saccharification is accomplished. The first step in the production of crystalline glucose, HFCS, fructose, and maltose is the production of glucose syrup. Thermostable α-amylase liquefies starch paste, primarily made of corn starch, which is then saccharified by fungal glucoamylase and debranching enzymes. As the affinity of glucoamylase reduces with large levels of dextrinized products, the efficiency of the saccharification process is dependent on the level of dextrinization during liquefaction. The solution is filtered using ion exchange treatment using cationic, anionic, and mixed resin beds after being converted to glucose, and then it is decolored. The process of evaporation is used to raise the concentration of glucose syrup. After the entire content of the glucose syrup has solidified or crystallized, the production of glucose powder and anhydrous dextrose begins. Because of its non-hygroscopic nature and low tendency to crystallize, maltose is mostly utilized as a sweetener in the food industry. It is used as an intravenous sugar supplement in the medical industry. For producing maltose, a variety of natural foods are used, including the starches of corn, potatoes, sweet potatoes, and cassava. For the production of maltose of the medical grade, a concentration of 10% to 20% is necessary, whereas, for the production of maltose of the food grade, a concentration of 20% to 40% is appropriate (Chae et al., 2010). Chocolate syrup is made from cocoa using alpha-amylase. Chocolate syrup, a stable low-viscosity substance with a flavor of cocoa, is used in the production of frozen chocolate and other flavor-infused confections. Amylase is used in the baking industry to hydrolyze starch in the dough. The fermentable by-products of α-amylase function as anti-staling agents and counteract the consequences of staling, such as loss of crust crispness and moisture content (Farooq et al., 2021). To increase the digestibility of fiber, amylases are also utilized in the clarification of beer, fruit juices, and animal feed production. Starch paste is used as an agent during weaving in the textile industry. The starch shields the string during warping from abrasion, cutting, and the buildup of static charge. Before scouring and dying, starch is removed from the fabric after it has been made. The most efficient method for improving the quality of the fabric is to utilize thermostable amylase in the enzymatic desizing process. Gelatinized starch was applied as a sizer to the fabric, which was then processed at a high temperature (90°C), treated with thermostable amylase, washed in hot water, then washed in cold water, and dried (Martinez et al., 2018).

3.9 ETHANOL

Ethanol is utilized as a biofuel to combat environmental pollution and global warming. Starch is employed as a substrate for this purpose because of its high availability and low cost (Wu et al., 2004). The process of making ethanol from starch involves steps. Thermostable amylase mediates the liquefaction of starch. Other sets of enzymes, such as amylase, glucoamylase, pullulanase, etc., saccharify liquid starch. Starch can be found in the waste products of several sectors, including textile, paper, sugar, bakery, confectionery, and brewing. Industrial wastewater clean-up is a costly operation that might occasionally expose the environment to biological risks. Enzymatic waste treatment is both environmentally and economically advantageous. Amylase has been used to treat starch waste so that it can be broken down and then turned into microbial biomass or biofuel. The bioconversion of wastes into hydrogen and methane can also be done via eco-biotechnological methods. After producing paper, the starch paste is used as a sizing agent to prevent mechanical damage during processing and lessen the tendency for liquid absorption when drying. Low viscosity starch solution that has been heated and treated with amylase is used as a sizer. This procedure improves the produced paper quality, texture, strength, smoothness, writing, and erasability capabilities. Ninety percent of liquid detergents have enzymes as their main ingredient. The use of enzyme-based detergent is safe for the environment. Amylase is the main component used to create enzyme-based detergents out of all the other enzymes. The starchy components are broken down by amylase to make the fabric whiter and brighter. Similar to this, used food dishes often have leftovers of starchy meal ingredients. The enzyme amylase eliminates these leftover starchy molecules.

Figure 3.6. Aspergillus oryzae.

*(**Source:** https://upload.wikimedia.org/wikipedia/commons/c/c9/Aspergillus_oryzae_%28%E9%BA%B9%29.jpg, Creative Commons License)*

As a digestive aid for the treatment of indigestion, amylase from Aspergillus oryzae (Figure 3.6) is marketed under the trade name Takadiastase(Zhu & Tramper, 2013). There are digestive medicine formulations in liquid and tablet form on the market. To create an ultrasensitive amylase biosensor, polysaccharides like glycogen and amylopectin were coated onto gold-coated quartz crystals or maltopentaose and β-glucosidase was immobilized on a membrane. For the diagnosis of acute pancreatitis, which is linked to hyperamylasemia and urinary amylase, this biosensor can track variations in the concentration of serum or urine amylase. Another source of serum amylase is biliopancreatic duct blockage. Amylase is also used as a biomarker for conditions such as stomach aspiration, acute stress, and pancreatic fistula. The diagnostic method for hyperamylasemia is therefore biosensor-mediated amylase detection. Treatment for postprandial hyperglycemia, obesity, and type 2 diabetes commonly involves the extraction of therapeutic inhibitors of amylase like flavonoids, alkaloids from wheat seed, white kidney bean, Phaseolus vulgaris, Sophora japonica, Varthemiaiphionoides, and Ginkgo biloba (Abd Elkader et al., 2022). The enzyme amylase is used for a variety of things, including in the food, biofuel, detergent, textile, and paper sectors. Bacillus licheniformis or Bacillus stearothermophilus amylase is employed in the sugar industries for starch hydrolysis in the starch liquefaction process, which is the foundation for the manufacturing of glucose syrups. Gelatinization, liquefaction, and saccharification are all steps in the enzymatic conversion of starch to glucose. The process of saccharification, which comes after hydrolysis, results in the production of glucose and maltose. The bread quality is increased when amylase is added to the dough. The baking business uses Bacillus stearothermophilus amylase on an industrial scale. Starch is the most used substrate for the manufacture of bioethanol in the biofuel sectors. To increase the writing quality of paper, the paper industries use cold active amylase, which lowers the viscosity of starch for proper coating of paper. This enzyme has uses in industries including medicine and the environment. The digestive enzyme amylase can be utilized to treat various digestive issues. Large-scale production of starch processing waste results in pollution, yet products contaminated with starch can be cleaned up using amylolytic enzymes found in microorganisms.

The amylase proteins from Aspergillus, Bacillus, and other genomes were aligned using the multiple sequence alignment (MSA) methods in BioEdit v7.2.5 (http://www.mbio.ncsu.edu/BioEdit/bioedit.html) using the default parameters (Kong et al., 2014). MEGA V6.06 was used to import the alignment file and create an unrooted Neighbour-Joining phylogenetic tree. The phylogenetic analysis employed only the aligned locations, leaving out less conserved and gap regions. With 1000 replications, bootstrap analysis was used to test the phylogenetic trees. When compared to other organisms like Bacillus amyloliquefaciens, GeoBacillus stearothermophilus, Bacillus subtilis, and Bacillus KSM (Prokaryote), previous studies on the evolution of amylase showed that amylases of Bacillus subtilis (Prokaryote) and Aspergillus oryzae (Eukaryote) are the closest. Synonymous codons are those that encode the same amino acids. Now, the synonymous codons may be employed to encode the amino acid in a different order. The preference pattern for Aspergillus is unclear, but it is evident that A or U-ending codons are preferred over G or C-ended codons in Bacillus. For instance, it is unclear what codon preference pattern applies to the amino acids alanine, proline, arginine, etc. Except for the codons CCG and ACG, which encode proline and threonine, respectively, the GC content pattern is entirely reflected in the pattern of codon usage of the gene encoding amylase in various species of Aspergillus and Bacillus (Xie et al., 2014). Even though the Bacillus gene for amylase has a poor GC content, these two codons are used more frequently in Bacillus than in Aspergillus. Twenty amino acids are encoded by 61 codons, meaning that more than one codon may encode a single amino acid. They are referred to as synonymous codons.

Calcium (Ca^{2+}) ions have a strong affinity for binding a variety of intracellular and extracellular proteins, such as enzymes, and structural and transport proteins. Ca^{2+} binding is a significant biochemical process with numerous effects. Up to nine ligands can be discovered in crystalline complexes involving calcium-carboxylate and calcium-water interactions, despite their propensity for octahedral coordination. It is vital to provide a detailed structural description of the pathways to comprehend how Ca^{2+} can affect protein activities.

The stability of the proteins is greatly enhanced by pure structural Ca^{2+} sites. Recent developments in protein engineering aim to create more reliable enzymes that may be used in demanding industrial settings. Designing new, superior enzymes will be made easier with a better understanding of how metal binding produces thermostability. Numerous intracellular functions have been linked to calcium, making careful regulation of Ca^{2+} concentrations necessary. The Ca^{2+} ions are coordinated by ligands inside the loop in the helix-loop-helix structure of the EF- motif. It has also been discovered that several additional proteins can bind with calcium. Ca^{2+} binding, for instance, orients nearby side chains in concavalin A in a way that is conducive to Mg^{2+} binding. A bacterial galactose-binding protein was also shown to contain Ca^{2+}. The stereochemistry of the two binding sites is close althought this periplasmic protein and the EF-type eukaryotic proteins have no evolutionary connection. In addition to phospholipase A2, it is found in a wide range of proteinases, including trypsin, subtilisin, thermolysin, and acid proteinase (Kovacs et al., 1988).

3.10 AMYLASES

Amylases have been found to have a Ca^{2+} ion that is tightly bound and necessary for the catalytic activity of the enzyme. Although these enzymes play significant biological roles and have potential industrial applications, little is known about their precise structure and catalytic mechanism. It consists of a cluster of residues from the larger domain A and the smaller domain B. Ca^{2+} links these two segments to maintain the structural integrity of the active site, which is thought to be a significant cleft between the ß barrel and the B domain (Schindelin, Marahiel, & Heinemann, 1993). In contrast, a single 12-residue-long region of a polypeptide chain forms the EF site. The stereochemistry of ligand coordination is the second peculiar characteristic. Ca^{2+} can bind up to nine ligands in organic complexes, but it prefers seven or eight. Though with different degrees of distortion, the Ca^{2+} is in octahedral coordination. Bacillus licheniformis amylase is extensively researched because of its excellent thermostability and practical value. An 8-barrel (TIM-barrel), the N-terminus, and the active site make up Domain A. Domain B is exclusively constituted by the β-strands with the largest structural diversity, while domain C, which forms a key motif and is situated on the opposite side of the TIM-barrel and contains the C-terminus. Bacillus amylase is stabilized and activated by calcium ions. Since the preserved Ca^{2+} is far away from the active site to directly engage in catalysis, it is thought to have structural importance. All amylase structures that have been discovered have the same calcium-binding site, which is situated right next to the active site at the junction of domains A and B. B. licheniformis and B. stearothermophillus amylases contain two more calcium-binding sites. The calcium ions CaI and CaII are situated close to one another and together with Na^{+} form a trio. There is still another calcium-binding site (CaIII), which functions as a bridge ion (Buisson et al., 1987).

The region around CaI and CaII in the amylase of B. licheniformis exhibits a triadic metal symmetry, and mutations affecting this region directly or indirectly are highly detrimental to the amylase activity at moderate temperatures. The position of glutamic acid residues within or close to the active sites of certain enzymes may influence their catalytic activity in addition to their multiple structural functions and their involvement in the regulation of other channels. The multidomain hydrolases and transferases of the amylase family require glutamic acid and aspartic acid residues to function. In the beverage and food sectors, food processing techniques have been employed to alter the food product and make it suitable for ingestion; these techniques either utilize chemical mediators or biological mediators. Since 6000 BC, enzymes have been utilized in processing techniques for manufacturing cheese, wine, bread, and beer. Enzymes are now widely recognized as useful products in the biotechnology sector. Enzymes have contributed to the textile, leather, detergent, pulp, and paper industries in addition to the food business. According to government regulations, food enzymes are primarily divided into two categories: food additives and food processing aids. The majority of food enzymes are employed as processing aids, but invertase and lysozyme are regarded as additives. Amylases hydrolyze the 1,4-glycosidic internal bonds of starch to create glucose, maltose, and maltotriose, which have molecular weights of 51.0–54.0 kDa. Amylases can be found in all kinds of animals, plants, and microorganisms. The three main genera for amylase production have been Bacillus, Penicillium, and Aspergillus. Compared to plant and animal amylases, microbial amylases offer a wider range of industrial applications (Saini et al., 2017). Numerous industrial uses for the amylase produced by B. licheniformis, B. amyloliquefaciens, and B. stearothermophilus include brewing, fermentation, food, paper, and textile industries.

Amylase is a protein with a molecular weight of 223.8 kDa that is used extensively in the sugar, starch, brewing, and beverage industries. This enzyme converts the substrates of starch and glycogen into maltose. This enzyme is produced by bacteria and fungi belonging to the Pseudomonas, Bacillus, and Clostridium families. Amylase from A. oryzae has been used for the production of food products such as organic acids such as citric acid, acetic acid, and soy sauce.

The 54 kDa microbial cellulase acts on the substrate of cellulose to produce monosaccharides. Although this enzyme has many uses in the manufacturing of pulp and paper, textiles, and biofuels, its main use is in the brewing business. The primary sources of this enzyme have been microbes, including bacteria, fungi, and protozoa. Cellulase principally consists of three sets of enzymes: exo-(1,4)-d-glucanase, endo-(1,4)-d-glucanase, and β-glucosidases. The first enzyme to be utilized domestically was pectinase. Pectinase can have a molecular weight of up to 31 kDa. It affects the pectic components in fruit juices, which clarifies the fluids as a result. It has also been for the manufacture of commercial wines and fruit juices. The primary microbial source for pectinase synthesis has been Penicillium sp. With a molecular weight of 205 kDa, invertase is neither poisonous nor pathogenic. The primary microbiological source for its production under controlled submerged fermentation has been Saccharomyces cerevisiae. It produces glucose and fructose from the sucrose substrate. This enzyme can be used in the manufacture of confections and the generation of ethanol from cane molasses (Figure 3.7).

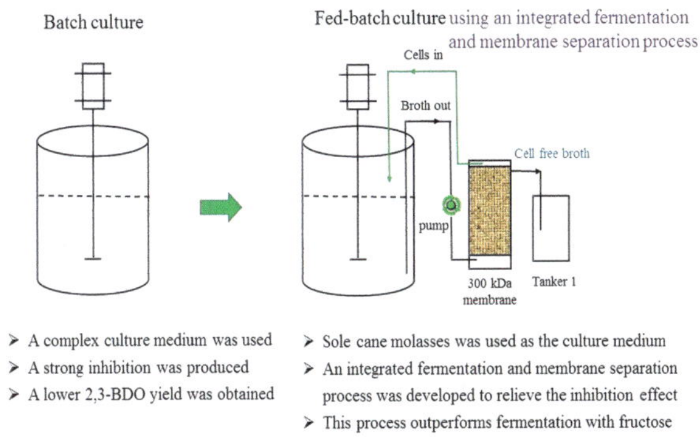

Figure 3.7. Cane molasses.

(**Source:** *Yang Y, Deng T, Cao W, Shen F, Liu S, Zhang J, Liang X, Wan Y. Effectively Converting Cane Molasses into 2,3-Butanediol Using Clostridium-ljungdahlii by an Integrated Fermentation and Membrane Separation Process. Molecules. 2022; 27(3):954. https://doi.org/10.3390/molecules27030954, Creative Commons License)*

3.11 PROTEASES

Proteases have a molecular weight of 126 kDa (Laan et al., 1988). It has a proteolytic system that breaks down protein peptide linkages into amino acids. Proteases are divided into endopeptidases and exopeptidases based on how they cleave peptide bond chains. Proline-specific peptidase enzymes, endopeptidases, aminopeptidases, and serine proteases are included in the category of proteases. All fungi and bacteria, but particularly Lactobacillus helveticus, L. acidophillus, L. bulgaricus, L. diebrueckii, L. lactis, and L. plantarum, have been used to acquire this enzyme. The enzyme has been extensively employed in food preparation because of its proteolytic properties. It has many applications in the food, detergent, and pharmaceutical industries. Proteases were extensively used in the dairy business. A 33 kDa water-soluble enzyme called lipase is employed to hydrolyze ester linkages found in lipid substrates into fatty acids. From Pseudomonas aeruginosa, Bacillus subtilis,

Staphylococcus aureus, Serratia marcescens, and other microorganisms, lipases or triacylglycerol acyl-hydrolases have been isolated. Free fatty acids, glycerols, esters, and glycerides are catalyzed by lipase to form esterified oils, which are widely employed in the pharmaceutical, chemical, and food industries. Lipase offers rancidity, little bitterness, and a distinct cheese flavor (Coulter & Combs, 1939). In the process of making cheese, lipases and proteases have made the cheese-less bitter and improved its flavor. The production of baked goods, fermented vegetables, dairy products, and fruit juices typically uses lipases. Three recombinant fungi from the Fusarium oxysporum, Thermomyceslanuginosus, and Rhizomucormiehi have been exploited in the food sector. Despite physical and geochemical conditions, some microorganisms, such as extremophiles, can endure harsh settings. Due to their nature, those bacteria are a more attractive enzymatic source than biological material, offering the highest level of stability. Although enzymes have proven very successful, they cannot be completely defined because isolated cultures of bacteria are not present. Since being extracted from acute conditions, proteases, lipases, amylases, cellulases, and other thermophilic enzymes are now utilized in a variety of industries.

Psychrophile-produced enzymes such as amylases, cellulases, lipases, and proteases have a variety of industrial uses, including the paper and pulp industry, the production of second-generation biofuel from lignocellulosic biomass saccharification and polymer-degrading activities. Particularly employed as additives in dishwashing and laundry detergents are proteases, lipases, and other thermos-alkaliphilic enzymes derived from bacteria that can survive under extreme pH levels. The cold climate of the alpine, polar, and ocean depths makes up about 85% of the earth's surface. These microbes routinely produce enzymes that are active at moderate to low temperatures. These enzymes exhibit strong catalytic activity at low temperatures but exhibit low thermostability at high temperatures. These enzymes have been used specifically for the processing of milk and fruit juice to prevent changes in nutrition and flavor at room temperature. Lactose-intolerant people can remove lactose from cold milk using the enzyme β-galactosidase (Saqib et al., 2017). The employment of cold-active lipases in the bakery as additives, cheese production, meat tenderization, and fermentation has made them an essential component of the modern food industry. Proteases are utilized in a variety of food-related sectors, including the production of cheese, baking, and brewing, as well as detergents, silk degumming, leather, and pharmaceuticals. Due to the lack of enduring stability, shelf life, recovery, and reusability of desired enzymes, as well as their potential applications, immobilizing enzymes has been the primary method of solving these problems. To improve the activity of enzymes, nanoparticles have been employed to immobilize and disseminate them in food matrices.

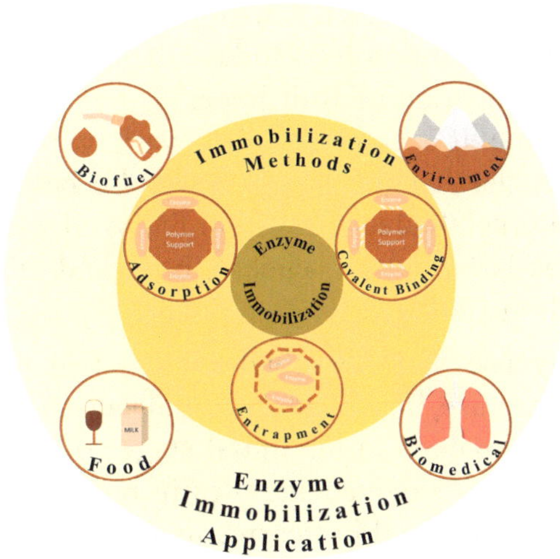

Figure 3.8. Immobilized enzymes.

Source: *Lyu X, Gonzalez R, Horton A, Li T. Immobilization of Enzymes by Polymeric Materials. Catalysts. 2021; 11(10):1211. https://doi.org/10.3390/catal11101211, Creative Commons License)*

3.12 IMMOBILIZED ENZYMES

Immobilized enzymes and their applications (Figure 3.8) in the food industry have grown gradually (Liese & Hilterhaus, 2013). With the aid of immobilized enzymes, which partially transformed fructose into gluconate, the starch was used to increase sweetness. The majority of enzymes used in the food business are those that break down and alter biomaterials to process food. Lipases have been used commercially to process beer, vegetables, baked goods, meat, and fruits as well as in the dairy industry for flavor development. For modifying the length of the fatty acid chains to improve the flavor of cheese, dairy industry lipases from A. oryzae, A.niger, and M. miehei have been employed. The important suppliers of commercial enzymes in the modern era have been microorganisms. Pectinase enzyme, which is mostly employed in the beverage and food sectors, is produced by all microorganisms, including fungi and bacteria. Pectinases and associated enzymes are frequently used in a variety of sectors, including the treatment of vegetable wastes, tea and cocoa fermentation, and fruit ripening. Esterase and lipase, two lipolytic enzymes, have been utilized in the breakdown of lipids. In the food business, cellulase has also been utilized to extract some compounds from soy protein, green tea, aromatic goods, essential oils, and sweet potato starch. In addition to cellulase, hemicellulase, and pectinase were used in the clarification of fruit juices. Amylase has been produced as an industrial enzyme. It is employed for a variety of purposes, including those in the bakery, starch liquefaction, alcoholic beverages, and animal feeds. Protease is a crucial enzyme used in the synthesis of various Asian delicacies as well as baked goods, cheese, beer, and other dairy products (Paulose & Chakraborty, 2021).

There are many uses for microbial enzymes in the dairy business, including protease, esterase, lipase, catalase, and lactase. Rennet, also known as rennin, has been employed in the initial stages of cheese production to act as a milk coagulant. To reduce allergies in products made from cow's milk, protease has been utilized in the ripening of cheese and modification of milk proteins. To enhance the lipolytic flavors of cheese, lipase has also been added during the ripening process. Lactase has been used to increase the sweetness and solubility of dairy products. By breaking up peptides, peptidase preserved the original flavor. The baking business has only employed amylase to make cakes and to make starch syrups, fruit juices, and digestive formulations. Amylase is typically utilized in the fermentation of flour because it breaks down the starch needed to make bread. Adding more sweetness has also improved the bread's flavor, taste, and color of the crust when it is toasted. The industry that makes bread has utilized xylanase and lipase to extend bread's shelf life. Food microbiology research is beginning to focus on fermentation technologies. Different fermented foods can be safely prepared via microbial interaction. Food technologists pay attention to indigenous fermentation processes to adopt solutions for food security. Many beverages and fermented foods have been made using lactic acid bacteria (LAB) and Saccharomyces cerevisiae. LAB are essential to the dairy production of acid and flavor (Menezes et al., 2018). Acetic acid bacteria (AAB) have been crucial in the production of a variety of alcoholic drinks and fermented foods like water kefir, cocoa, kombucha, lambic beer, and lamb.

The metabolism of cholesterol and inhibition of the angiotensin-converting enzyme is linked to regular consumption of kefir (Quiros et al., 2005). It has been linked to antibacterial activity, tumor suppression, wound healing, and immune system strengthening, which reduces allergy and asthma symptoms. In the southwest of Nigeria, fufu, ogi, and gari have been used as fantastic probiotic sources to cure a variety of gastrointestinal (GI) diseases like diarrhea and dysentery. In the countries of Africa, a variety of traditional fermented dishes are prepared using a common substrate of cereals, dairy products, fish, fruit, legumes, meat, root crops, and vegetables. In various European nations, low and non-alcoholic fermented beverages such as hardaliye, kefir, kombucha, and kvass are typical traditional fermented foods. Staphylococcus carnosus and S. xylosus were utilized as the starter cultures for the delicious fermented meat product known as jerked beef or charqui (Pinto et al., 2002). For rural residents who are below the poverty line, fermented foods are excellent nutritional supplements. The majority of rural residents of Emene, Nigeria, from infants to adults, regularly consume fermented foods to complement their diets with vitamins, proteins, minerals, and organic acids. By producing and selling various fermented foods and beverages, the socioeconomic standing of individuals in developing nations has been rising. Decarboxylation of the pyruvate results in the formation of the two-carbon molecule acetaldehyde, which is then reduced to yield ethanol. One glucose molecule is split into two molecules of ethanol and carbon dioxide during the process.

Acetobacter and Gluconacetobacterium have been used to produce vinegar in commercial quantities. In this process, ethanol is fermented in aerobic conditions to produce acetic acid. In acetone-butanol-ethanol (ABE) fermentation processes using lignocellulosic materials as a substrate on an industrial scale, Clostridium acetobutylicum is a useful bacteria (Sarangi & Nanda, 2018). The best substrate for the fermentation of propionic acid has been lactate. The bacteria Propionibacterium acidipropionici is used in this process. Acetate, formate, lactate, succinate, and ethanol are produced during a type of anaerobic fermentation known as mixed acid fermentation together with CO_2 and H_2. The process of fermentation is a biotechnology approach used in the food industry. This process involves the production of metabolites by microorganisms in an artificial setting. By offering ideal conditions for population growth, the incubation system used is intended to maximize the effectiveness of the target microorganisms. Several fermented dishes have been made using cereals, fish, fruits, milk, meat, and vegetables. Beer, rum, vodka, whisky, and other alcoholic beverages include ethyl alcohol, which is produced during the alcoholic fermentation process by yeast and other fungi. Bamboo leaves are used to cover the stem top, which is then allowed to ferment with Lactobacillus brevis, Lactobacillus plantarum, and Pediococcuspentosaceus for 7–15 days in an acidic environment. A traditional fermented meal in the Indian state of Manipur is called hawaijar (Jeyaram et al., 2008). After a thorough washing, soybean seeds are boiled for 2–3 hours. It takes 40°C to allow soaked seeds to cool. Then it is layered alternately with fresh banana or fig leaves in a bamboo basket. Bacillus subtilis, B. licheniformis, B. cereus, Staphylococcus aureus, S. sciuri, and Alkaligenes sp. are added to the vessel, which is then incubated at 25°C–40°C for three to five days. Several substrates are used to produce fermented foods, food additives, and beverages. Food additives raise the nutritional value of food. The fermentation process involves a single bacterium or many different microorganisms. Some wild and locally accessible fruits and vegetables have been used by ethnic communities as a base for making traditional fermented dishes and drinks.

CHAPTER

4 Fermentation Technology

Contents

4.1 INTRODUCTION

Fermented foods and drinks existed long before microbiomes, genome sequences, starter cultures, etc. Fermentation was just a natural process that happened when raw food items were left in an unprotected state. It took centuries for people to learn how to manage environmental factors so that fermented food products could be regularly produced at a high standard. The methods for making fermented foods seem to have separately evolved on every continent and are only based on empirical data. Egyptian bakeries and breweries were already mass-producing bread and beer as early as 3000 to 4000 BCE (Tamang et al., 2020). Historical records show that the manufacture and consumption of wine and other fermented foods and beverages occurred at the same time that civilizations spread over the Middle East, Europe, and the Mediterranean. It is interesting to observe how different fermented food consumption was in China, Japan, and other parts of the Far East from the Middle East. Instead of using animal products, the primary raw components were plants and shellfish. It is evident that these fermentation procedures in Asian countries developed and were widely used during the same period. Early studies on the origins of food fermentations tended to concentrate on alcoholic fermentations, specifically wine and beer. Researchers relied on chemical and molecular data from the artifacts found in various places throughout the world.

They used an approach that specifically involves collecting residues from prehistoric clay pottery jars and vessels discovered in sites that had been excavated, primarily in the East and China. Since these containers typically feature pores, any organic material would have been adsorbed and confined inside the pores of the vessel. This substance would have been resistant to microbiological or chemical deterioration in a dehydrated state. To determine the general age of these vessels, carbon dating is performed. The chemical contents are then identified using a variety of analytical approaches, such as gas chromatography, mass spectrometry, Fourier transform infrared spectrometry and other methods. The tests found many components, including tartaric acid, which is typically absent elsewhere but is found in high proportions in grapes and wine. Wine production may have started as early as the Neolithic Period (8500 to 4000 BCE) in the eastern regions near modern-day Turkey, Egypt, and Iran (Cavalieri et al., 2003). Two liver enzymes, alcohol dehydrogenase (ADH) and aldehyde dehydrogenase, normally metabolize ethanol. Alcohol poisoning will develop from even minute amounts of alcohol drinking if one of these enzymes is missing. For those who express these enzymes, acetaldehyde, a harmful intermediate, builds up in the liver if more ethanol than the liver can metabolize is consumed. Expression of a functional ADH was caused by a mutation in a particular ADH gene (ADH_4). Mutation shappened millions of years ago and they coincided with the emergence of ethanol-containing foods in nature.

Primates who ventured from the safety of the trees to consume the fruit that had fallen and who could withstand alcohol had a selective nutritional advantage. Researchers concluded that long before hominids appeared and drank alcohol on purpose, they had mastered the skill of managing their drinking (alcohol). The development of brewing, winemaking, and other ethanol-related technologies was ultimately influenced by our ancestors' ability to handle ethanol. As new raw materials, technology, and cuisines were adopted from fermented foods spread throughout the world, fermentation became an even more common practice during the Graeco-Roman era. These meals were crucial for far-flung armies and navies as well because of their improved storage stability and, in some cases, ease of transport. Cheese would have been a more dependable and condensed supply of nutrients than milk. People frequently chose beer and wine over water. The ability to conduct trade advanced during the Roman Empire. As a result, wheat for producing bread and cheese was made available throughout the Mediterranean, Europe, and the British Isles. Even though some societies produced bread in ancient Egypt, during the Middle Ages, the production of cheese, beer, and other goods fell under the purview of skilled artisans. These abilities were frequently taught to the following generation after being acquired. These artisans were frequently monks working out of monasteries and churches for some products, particularly beer and wine; this practice has a long history. In reality, a lot of the manufacturing techniques used today were first created by monks. For instance, 300 years ago, Benedictine monk Dom Pierre Perignon is credited with developing the method used to create carbonation in Champagne. Over time, the production of these goods grew more privatized, while frequently still subject to some degree of state regulation, which conveniently made taxes possible.

4.2 FERMENTED FOODS

Fermented foods have been among the important cultural and dietary foods ingested by humans from the Neolithic Period to the Middle Ages to the present. It is possible to argue that fermented foods and the ensuing technological advancements for their manufacture significantly influenced the dietary, cultural, and social evolution of humans throughout history. Nearly all civilizations and cultures rely heavily on fermented foods for their diets and cuisines, and many of these foods and beverages are consumed as part of religious practices, rites, and rituals (Tamang, 2010). The fermented foods sector of the food processing business has undergone significant transformation over the past few decades, much like other areas of the sector. The average size of a typical production plant has multiplied many times, and with it has the throughput i.e., the speed at which raw materials are transformed into completed products. They are widespread in Europe where several different fermented foods frequently call for or enforce traditional manufacturing practices. In the past century, the fundamental method by which fermented foods are made has evolved along with the size and scope of the sector. Pasteurized milk, which satisfies stringent quality criteria, is now more widely used, at least in the US, Canada, New Zealand, Australia, and portions of Europe. Although there are still some open, copper-lined cheese containers in use, most current manufacturing tanks and containers are covered and made of stainless steel or other materials that make cleaning and sterilizing procedures easier. Modern facilities are specially designed with an emphasis on sanitary requirements so that exposure to bacteriophage and airborne microbes is reduced. To avoid unintentional cross-contamination, the entire plant architecture, air management, and personnel access are taken into account.

Even though the container sizes were small, conventional production techniques may not have consistently produced goods with good quality features. The cost of raw materials alone for a day's worth of production might be thousands of dollars. A producer simply could not absorb or put up with such losses, even if they only happened occasionally. Particularly, food safety has gained international attention. For the fermented food sector, reducing the risks connected to food-borne viruses or other food safety concerns is a top priority. There may be a zero-tolerance policy for infections and other dangers in fermented foods, depending on the location. There are currently quality assurance processes in place in the food sector, with the main objective being to consistently produce safe food items. The fermented food industry has essentially changed from being primarily an art or craft-based profession to one that depends on contemporary science and technology. All processed foods, not just fermented foods, must adhere to strict standards of hygiene, quality, and consistency. The production of fermented foods depends on biological elements, especially the cultures, which sets them apart from other types of food. It is the only sector of the food processing industry where the existence, development, and activity of microorganisms are necessary for the success of the product. The fermentation may not succeed, resulting in decreased product quality and safety if a certain attribute is lost, the cells become less viable, the culture is infected by viruses, or the culture is impeded by other environmental conditions. Although salt and other anti-microbial substances can be applied, biological means are generally used to control diseases. The implications are therefore quite important.

4.3 MICROORGANISMS

Microorganisms that start fermentations are more difficult to standardize than favoring, nutritional, or functional food additives. This is due to the possibility of variations in their viability, biochemical activity, and even concentration from batch to batch or during storage. Even though there are commercially available starting cultures with standardized cell counts and activity, many industrial fermentations still depend on the presence of naturally occurring microbes. This local microbiota composition and biological activities are frequently prone to variations. Bacteria are frequently exposed to a range of biological and chemical inhibitors in food and the environment, which can impair their viability and activity. It is hard to estimate the importance of culture organisms in guaranteeing the safety and quality of food because they are frequently the primary method used to reduce spoilage and pathogenic microbes in fermented foods, whether they are introduced or native. The completed product will be vulnerable to spoilage, quality errors, or worse if the culture is unable to function efficiently. Therefore, the problem facing the fermented food sector is to produce products whose production is already susceptible to intrinsic unpredictability while still meeting criteria for consistency, quality, throughput, and safety in the modern period. Foods that had undergone fermentation were seen to be highly preserved, at least when compared to their raw ingredients. Particularly in temperate climates, raw food material like milk or meat had to be consumed promptly after it was collected because it would soon expire. Although some items may have been preserved by drying, smoking, or salting, fermentation must have been a preferable alternative because of some desired qualities. One of the primary reasons fermented foods became so prevalent in human diets and contributed to practically every cuisine on every continent was preservation. Fermented foods still have crucial preservation and extended shelf-life qualities (Obafemi et al., 2022).

The majority of fermented foods are made from nutrient-dense ingredients like milk, pork, rice, soybeans, and wheat. Fermented foods have long been thought to provide higher nutritional value, although many of these claims have lacked a solid scientific foundation. There is proof that many fermented goods have improved nutritional qualities as a result of fermentation. Yogurt food makes an interesting case from a nutritional standpoint. Because the majority of people in the world are unable to manufacture the enzyme galactosidase, which is required for the digestion of lactose, the primary sugar found in milk, fluid milk is not widely consumed. When people with low levels of galactosidase synthesis consume lactose, a physiological condition known as lactose malabsorption/intolerance results. This condition is accompanied by some unpleasant but non-life-threatening symptoms. These include cramping, bloating, diarrhea, and mild to severe digestive distress. Although adult Caucasians may also be lactose intolerant and many people from these countries consume dairy products without experiencing any negative effects, these populations are particularly prone to this problem, notably those from Asia, Africa, and the Middle East (Deng et al., 2015). It's interesting to note that multiple studies have shown that those who are lactose intolerant may drink yogurt without experiencing any negative side effects and thus can benefit from the nutritional benefits of milk, such as calcium, high-quality protein, and B vitamins.

Another excellent example of a fermented food article that seems to have better nutritional value than the source material it was made from is wine. There is currently a wealth of research demonstrating that wine includes substances other than ethanol that may improve health. Red wine contains resveratrol, among other phenolic compounds, which have been found to have antioxidant properties that may lower the risk of cancer and heart disease. In the Mediterranean, where heart disease death rates are often lower than in other areas, wine and other fermented foods are extensively consumed. It has been proposed that a Mediterranean diet would be the best for human health (Bonaccio et al., 2012). It has been hypothesized that eating foods that have undergone fermentation may provide health benefits that go beyond the macronutrients or metabolic by-products created by the fermentative microbes. In particular, it is currently believed that the bacteria responsible for the actual fermentation support human health, particularly in the gastrointestinal system. Attention has been paid to yogurt and other cultured dairy products as sources of beneficial bacteria. Several fermented foods, such as fermented vegetables, grains, and soy products, may also contain beneficial microorganisms. Consuming fermented foods has been linked to even extraintestinal advantages, such as improvements in mental health, cognition, and sharpness (Casertano et al., 2022). It is now advised to consume fermented foods as part of a healthy diet because their nutritional value has been recognized.

4.4 NUTRITION

The fermented foods are nutritionally different from the ingredients used to make them. They differ significantly in terms of their functional characteristics. For instance, the appearance and flavor of cheese differ from those of milk. The functional improvement in bread and beer is likely the most obvious. When people first started using wheat as nourishment some 10,000 years ago, there wasn't much they could do with the flour other than produce basic dough (Faris, 2014). However, letting the dough ferment before baking would turn the flour into tasty bread. Another widely utilized grain was barley, which was frequently used in dishes. However, barley's low gluten concentration compared to wheat limited its use in bread. The importance of barley in beer, a fermented beverage, has increased. Beer is the best example of functionality that has been improved by fermentation because barley is the primary ingredient, aside from water in its production. Fermented foods were likely first consumed and spread around the world due to preservation, nutrition, and functioning, but there was another crucial factor that contributed to their popularity. Foods that had undergone fermentation must have tasted, smelled, and looked vastly different from their original raw forms. Many of these contrasts would have been appropriately regarded both then and now. There is a debate that fermented foods have aroma, flavor and other visible qualities that are quite unlike the raw materials from which they are made. There isn't a technique to manufacture fermented foods without the fermentation process, with a few exceptions. The rich flavors in Swiss and Gouda cheeses, which are generated through fermentation alone, cannot be produced mechanically or by any other method.

Similar to kimchi, dry-cured salami, (Figure 4.1) Champagne, Pilsner beer, and balsamic vinegar, none of these products can be made any other way. Even the manufacturing processes for many fermented items are distinct and call for rigorous inspections. Parmesan cheese must be produced in a certain area of Italy, using milk from only certain breeds of cows, by time-honored practices, and aged under specific conditions. The final product must adhere to quality standards.

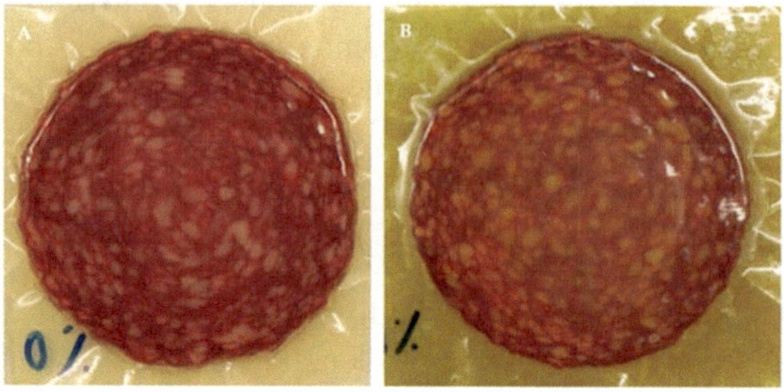

Figure 4.1. Dry-cured salami.

(**Source:** *Andrade MA, Ribeiro-Santos R, Guerra M, Sanches-Silva A. Evaluation of the Oxidative Status of Salami Packaged with an Active Whey Protein Film. Foods. 2019; 8(9):387. https://doi.org/10.3390/foods8090387, Creative Commons License)*

The name Parmesan is forfeited in the event of any deviation. However, fermented items can be produced without fermentation, including several types of fresh cheese, sausages, and even soy sauce. Acids, enzymes, or tastes are typically added directly to the product during production to mimic the functions that fermentative microorganisms ordinarily carry out. They typically lack the flavor and general organoleptic qualities of their traditional fermented equivalents. As the potential positive effects of fermented foods on human health become more thoroughly demonstrated, both scientifically and clinically, consumption of these items will probably rise as well. There is strong evidence to support the claims that living bacteria contained in cultured milk and other products, as well as red wine, may lower the risk of heart disease and improve gastrointestinal health. The capacity to get genomic data on the microbes responsible for food fermentation may be a substantial advancement in the last ten years. Today, it is possible to measure and describe the composition of entire microbial communities swiftly and affordably. It is also possible to evaluate the functional significance of particular species participating in fermentations. Similar to this, metabolomics has developed to the point where it is now possible to identify and measure the metabolic by-products of fermentation, even those produced at extremely low levels. It is now possible to relate the metabolic contributions of specific microbes or groups of bacteria to their organoleptic contributions. Researchers may have the opportunity to custom-produce fermented foods that not only have predictable flavor and other functional qualities but also give consumers nutritional benefits (Gao et al., 2021).

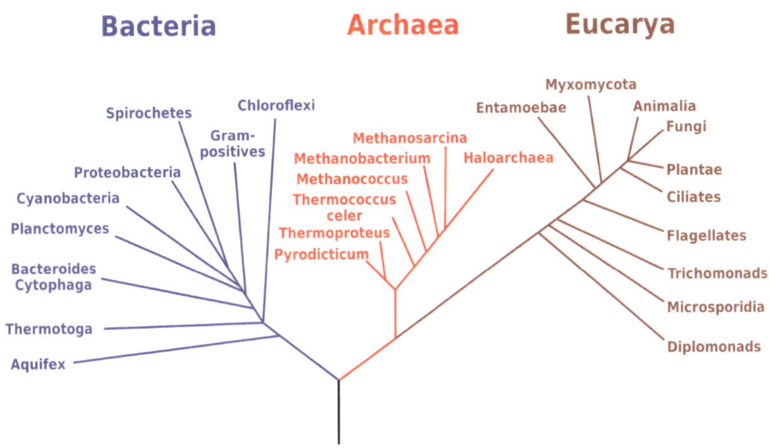

Figure 4.2. Eukarya.

(**Source:** *Corsini M, Zanello P. The Redox Active [2Fe-2S] Clusters: Key-Components of a Plethora of Enzymatic Reactions—Part I: Archaea. Inorganics. 2022; 10(1):14. https://doi.org/10.3390/inorganics10010014, Creative Commons License)*

4.5 EUKARYA

The Eukarya (Figure 4.2) is the group of organisms that are most relevant to fermented foods. Despite being fascinating to biologists for a variety of reasons, the Archaea are rarely connected to food. Instead, they are made up of organisms that thrive in situations with extremes like extremely high temperatures, extremely low pH levels, or extremely high salt concentrations. However, some Archaea dwell in the gastrointestinal system and release methane; as a result, there is currently a lot of interest in how these organisms affect human health. Several traits determine whether an organism is eukaryotic or prokaryotic. For instance, more than 50 cytological, chemical, metabolic, molecular, and reproductive characteristics are included in The Ninth Edition of Bergey's Manual of systematic bacteriology (Garrity & Holt, 2001). The existence of a nuclear membrane in eukaryotes and the absence of one in prokaryotes, however, has traditionally been used to distinguish between eukaryotic and prokaryotic species. As a result, prokaryotes include bacteria and archaea. In contrast, the Eukarya domain includes fungi, protists, mammals, and plants. Fungal taxonomists have hypothesized that the Kingdom Fungi has somewhere between 1.5 and 5 million different species, while the actual number is unclear (Richards et al., 2017). Only 1 to 5 percent of fungal species are known at this time. In addition, there is no consensus among taxonomists over how to classify fungi into higher taxa. The fungal taxonomy also has a lot of names that are interchangeable and repetitive. Only 100,000 of the roughly 400,000 species names that have been proposed in the literature have been recognized by the current taxonomy. Eumycota members are classified into seven major groups, or phyla, and 10 subphyla according to the recent fungal taxonomy. There are up to 900 species in the Zygomycota category, however, those that are significant for foods and food fermentations are mostly found in the Rhizopus and Mucor genera (Malosso et al., 2006). Some strains in particular contain proteolytic enzymes that resemble chymosin, the enzyme used to coagulate milk during the production of cheese. Others employed in the fermentation of tempeh.

With up to half of all fungal species contained within it, the Ascomycota is a large phylum of fungi. Members of the Ascomycota known as Ascomycetes have septate mycelium with cross walls or septa and generate ascospores, which are sexual spores, inside an ascus. The ascomycetes create asexual reproductive states, in which conidia, or asexual spores, are produced and used for reproduction. From the tip of certain airborne viable hyphae, these spores are freely dispersed (conidiophores). Single-celled organisms known as yeasts are an important group of ascomycetes. Although yeasts can form ascospores inside their cells, they can also reproduce asexually through a process called budding. Because they can produce ethanol and carbon dioxide, yeasts are crucial to the fermentation of food and beverages. Organizing the Bacteria domain has been just as difficult as organizing the Fungi domain. The first sequenced genome of an organism engaged in food fermentation, Saccharomyces cerevisiae, was made public in 1996. Despite being a lab strain and being somewhat different from the ordinary baker's or brewer's yeasts used in industry, the sequencing strain nonetheless offered important insights into the biology and genetics of yeast. A few years later, researchers revealed the genome sequence of the lactic acid bacterium Lactococcus lactis subsp. lactis. Despite the wide variety of bacteria involved in the production of fermented foods, almost all of them are currently grouped into one of three phyla: Firmicutes, Proteobacteria, or Actinobacteria. The Firmicutes are the most important of them. The primary organisms employed in the production of fermented foods include the lactic acid bacteria, a collection of Gram-positive bacteria. Bacillus and Staphylococcus are two other genera that belong to this phylum. They are made up of species that are solely employed in a few specific fermented dishes. Gram-negative bacteria found in the Proteobacteria are crucial to the fermentation of vinegar. Actinobacteria, which also includes the genera Propionibacterium, Kocuria, Micrococcus, and Bifidobacterium, include other species that are related to fermented foods (Tamang et al., 2017). It's important to note that a variety of other microbes that accidentally contaminate fermented foods may also be present.

It is crucial to note that the lactic acid bacteria itself has no recognized standing in taxonomy. Instead, it serves as a convenient general phrase to define a collection of genetically and functionally related microorganisms. However, the term has a pretty significant connotation among microbiologists and those who study food fermentations, similar to coliforms or psychrotrophs. The term "lactic acid bacteria" refers to a group of Gram-positive, non-spore-forming, low-G+C, lactic acid-producing rods and cocci that share some biochemical, physiological, and genetic characteristics. Numerous morphological and genotypic distinctions set them apart from other Gram-positive bacteria (such as Bacillus, Listeria, and Bifidobacterium) that also make lactic acid. Many times, lactic acid bacteria are characterized as meticulous organisms with intricate dietary needs. There are species that, in ideal circumstances, will only flourish on nutrient-rich, well-fortified medium or food. There are, however, some lactic acid bacteria species that are extremely adaptable in terms of their environment of growth and can still thrive in less-than-ideal nutritional environments. Some lactic acid bacteria are known for their capacity to flourish in unfavorable conditions, including those that are frequently present in fermented foods. These organisms can be isolated from ethanolic environments, salt brines, low pH foods, and other sources besides just plant material, milk, and meat. This implies that, in addition to their beneficial effects, lactic acid bacteria can also occasionally lead to the deterioration of fermented foods. The characteristics of lactic acid bacteria that pertain to nutrition metabolism may be the most important ones. After all, the capacity of lactic acid bacteria to metabolize carbohydrates and produce lactic acid and other metabolic end products is one of the key reasons why they are utilized in fermented foods. More than 90% of the sugar substrate is solely converted to lactic acid in the homofermentative pathway (Y. Wang et al., 2012). Lactic acid, acetic acid, ethanol, and carbon dioxide are the four end products of the heterofermentative route. The lactic acid bacteria group has 40 genera according to current taxonomy, but only around 12 of them are thought to be important in fermented foods. Lactic acid bacteria can be placed in a large phylogenetic cluster close to other low G+C Gram-positive bacteria based on 16S rRNA sequencing and other molecular approaches. A Streptococcus-Lactococcus branch, a Lactobacillus branch, a distinct Lactobacillus-Pediococcus branch, an Oenococcus-Leuconostoc-Weissella branch, and a Carnobacterium-Enterococcus-Tetragenococcus-Vagococcus branch are the five subclusters that are dominant. It is important to note that this phylogeny does not completely match the physical and physiological traits of these bacteria. Even though Lactobacillus brevis and Pediococcuspentosaceus belong to the same subcluster, the former is homofermenting cocci while the latter are rod-shaped heterofermentative bacteria. Both Carnobacterium and Enterococcus are homofermentative cocci, and both are obligate heterofermentative rods.

4.6 PHYSIOLOGY

Even though there are just a few physiological differences between L. lactis subsp. lactis and L. lactis subsp. cremoris, at least two of these changes are important during milk fermentations. For instance, L. lactis subsp. lactis strains can grow up to 40°C, whereas the majority of L. lactis subsp. cremoris strains cannot grow above 38°C. Compared to L. lactis subsp. cremoris, L. lactis subsp. lactis has a higher tolerance to salt. These variations are significant since temperature and salt are two of the main factors used to regulate the activity of the starting culture when making cheese. L. lactis subsp. cremoris is thought to be less resistant to environmental constraints that are present throughout the growth of cultures and during dairy fermentations (O'Sullivan & Condon, 1997). Temperature and salt concentrations must be adjusted to accommodate for organism development requirements if L. lactis subsp. cremoris is employed as a culture for cheesemaking. The genera Oenococcus and Weissella, which are linked to one another and the Leuconostoc, are members of the Leuconostocaceae Family. Mesophilic Leuconostoc grows at temperatures between 18 and 25 °C. At temperatures below 10°C, several species can grow. Depending on the structure and chemistry of the growing media, they can appear coccoid or rod-like under a microscope. They lack a functional glycolytic pathway and depend on the phosphoketolase route to use carbohydrates. Leuconostocs are facultative in terms of environmental requirements because certain species grow better in an aerobic environment while others grow in an anaerobic environment. Leuconostoc species frequently contain plasmids and they encode for lactose and citrate metabolism and bacteriocin synthesis. Leuconostocoenos was earlier categorized as Oenococcusoeni(Dicks, Dellaglio, &Collins, 1995). Among the lactic acid bacteria, Weissella and Lactobacillus are important.

They can be discovered in sediments, lakes, soil, and organic matter. They have also been isolated from humans and other animals. Weissella is consequently linked to the urogenital tract, gastrointestinal tract, and oral cavity. Weissella species are foodborne and have frequently been found in fresh, rotten, and fermented foods. Certain strains produce bacteriocins, indicating the potential for these organisms to act as antimicrobial agents.

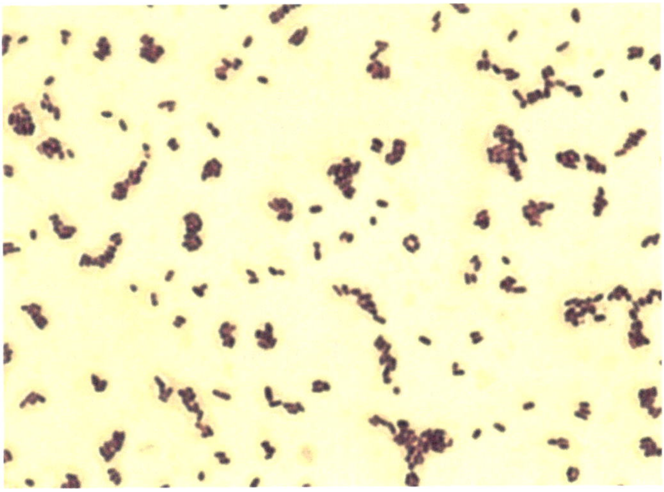

Figure 4.3. Weissellaconfusa.

(Source: https://ars.els-cdn.com/content/image/1-s2.0-S2590170221000157-gr1_lrg.jpg, Creative Commons License)

A few studies have suggested that Weissella, primarily Weissellaconfusa (Figure 4.3) and Weissellacibaria, may produce opportunistic infections, causing disease in both immunocompetent and immunocompromised people. The majority of these instances included non-life-threatening blood infections (septicemia). The isolated strains in the patients were influenced by the fact that they were typically vancomycin-resistant. The majority of species are linked to particular environments, such as plant and vegetable matter, milk and dairy, and meat products. Leuconostocgasicomitatum, for example, is one species that contributes to food deterioration, whereas other species play a key role in fermented foods. Leuconostocmesenteroides subsp. mesenteroides, Leuconostockimchii, and Leuconostocfallax, which are connected to vegetable fermentations, as well as Leuconostocmesenteroides subsp. cremoris and Leuconostoc lactis are employed in dairy fermentation. The specific sugars that these strains ferment generally reflect the ecosystems where they inhabit. While dairy strains like L. mesenteroides subsp. cremoris and L. lactis are more likely to ferment lactose, galactose, and glucose, plant-associated strains like L. mesenteroides subsp. mesenteroides ferment plant sugars i.e., fructose, sucrose, arabinose, and trehalose. Leuconostocs use heterofermentation to break down sugars into a variety of by-products, including lactic acid, acetic acid, ethanol, and carbon dioxide. These bacteria can be distinguished from homofermentative streptococci, lactococci, and pediococci using diagnostic tools. The pediococci are very similar to other coccoid-shaped, obligate homofermentative lactic acid bacteria. These bacteria divide into two planes that are perpendicular to one another. Tetrads are created as a result, which can be seen using a microscope. Cells may arise in pairs, but unlike lactococci, streptococci, and leuconostoc, no chains are produced. Like other lactic acid bacteria, pediococci are facultative anaerobes with intricate dietary needs.

Although some species can thrive at temperatures as high as 50°C, their ideal growth temperatures range from 25 to 40°C. The capacity of some pediococci to endure high acid and high salt environments differentiates them from other lactic acid bacteria. They belong to the family Lactobacillaceae and are closely connected to the Lactobacillus casei subcluster. The pediococci can be found in a variety of environments, including milk, brines, animal feces, and beer. Pediococcus presently has 12 identified species, many of which play significant roles in food fermentations. Two species, Pediococcusacidilactici and Pediococcuspentosaceus, are found in raw vegetables. Under the right conditions, these microorganisms are essential for the production of sauerkraut and other fermented vegetables. It is also possible to combine these species with pork to create fermented sausages. P. acidilactici and P. pentosaceus are occasionally detected in cheese, where they might take part in the ripening process despite not being able to ferment lactose. In particular, beer, wine, and cider include microbes called pediococci that are significant. One species, Pediococcusdamnosus, is particularly problematic in beer because it creates diacetyl, a significant problem in beer. This strain has been linked to the generation of biogenic amines, which suggests that it may also be responsible for wine deterioration. In pediococci, plasmids are frequently found. These plasmids contain genes that may code for processes related to the synthesis of bacteriocins and the metabolism of sugars like raffinose and sucrose. These are classified as antimicrobial proteins that inhibit a wide range of bacterial species, including Listeria monocytogenes, Staphylococcus aureus, and Clostridium botulinum, which are pathogens linked with meat. Since pediococci are utilized as starter cultures for the fermentation of sausage, bacteriocin synthesis is a particularly vital characteristic in pediococci. Therefore, when used in the production of sausage, bacteriocin-producing pediococci could offer an additional level of preservation.

4.7 LACTOBACILLUS

More than 250 species and subspecies of Lactobacillus are currently recognized (Jarocki et al., 2020). The only thing all of the species in this genus have in common is that they are all rods. Some species are very small, whereas others are longer. Therefore, it might be challenging to determine whether a particular strain is even a rod when examined under a microscope. Colony morphology varies among strains on agar plates, with some creating big, rounded colonies and others producing small ones. Although the G+C content of lactobacilli can range from 32 to 55 percent, the G+C content of lactic acid bacteria as a whole is typically less (35 to 40 percent). This genus of lactic acid bacteria is environmentally, physiologically, biochemically, and phylogenetically diverse, as well as having the greatest number of species and morphological variability. In terms of ecology, lactobacilli can be found in a variety of settings. Some species are present in plant and animal matter and can be found in dairy and meat, juice and fermented beverages, grains, and cereal foods. The idea that these bacteria contribute to intestinal and extraintestinal health has been supported by scientific evidence due to their prevalence in the gastrointestinal tracts of humans and animals, as well as in the stomach, mouth, and vagina. Lactobacilli are fermentative, like all lactic acid bacteria, but they also have a wider range of metabolic processes than other lactic acid bacteria. The species can be broadly divided based on the metabolic routes they employ to ferment sugars. Some lactobacilli can be specific despite their diverse metabolic capabilities, and many species need habitats that are rich in nutrients. They are not particularly lipolytic or proteolytic, and fatty acids, amino acids, and peptides are typically needed for rapid growth. The majority of lactobacilli also require carbohydrates that can be fermented, and depending on their source or environment, they can ferment a variety of sugars. Therefore, in addition to fermenting the majority of common sugars like glucose, fructose, and lactose, several plant-derived carbohydrates, including cellobiose, amygdalin, and trehalose, can also be used.

Many Lactobacillus species are important in fermented foods. Some strains are added directly in the form of starter cultures, but there are endogenous lactobacilli that are present in the raw materials and contribute slightly to the production of fermented foods. Dairy and sausage applications make up the majority of uses for starter culture lactobacillus (Kuleasan & Cakmakci, 2002). The two main species utilized as dairy starting cultures are Lactobacillus helveticus and Lactobacillus delbrueckii subsp. bulgaricus, mostly for cheese and yogurt. Other widespread species connected to dairy products include Lactobacillus acidophilus and L. casei. Lactobacillus plantarum and Lactobacillus sakei subsp. sakei are frequently found in starter cultures for sausage fermentation. Heterofermentative Lactobacillus sanfranciscensis, L. brevis, and other lactobacilli are used to make bread. While there are pure starting cultures for bread, wild cultures are still frequently employed. The natural microflora is required to start and complete the fermentation process for the production of sauerkraut, kimchi, and other fermented vegetables, though pure starter cultures containing L. plantarum, L. brevis, and related species have become more popular for controlled pickle fermentations.

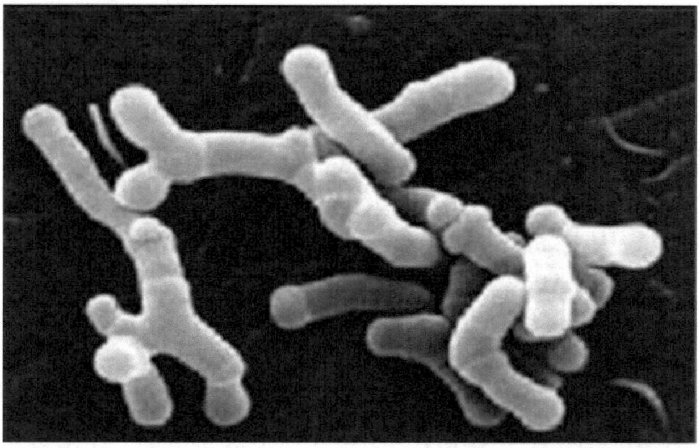

Figure 4.4. Bifidobacterium.

(**Source:** *https://upload.wikimedia.org/wikipedia/commons/4/41/Bifidobacterium_longum_en_microscopie_%C3%A9lectronique.jpg, Creative Commons License)*

It is debatable whether Bifidobacterium (Figure 4.4) species should be regarded as microbes that fermented food (Selhub et al., 2014). They do have a fermentation-based metabolism since they produce lactic and acetic acids. These bacteria are not used in the production of any fermented foods and are phylogenetically distinct from lactic acid bacteria. Bifidobacteria are discovered in raw food sources. However, because bifidobacteria are added to diets as probiotics to provide health advantages, they are crucial in some foods, primarily fermented dairy products. Their major habitat is the intestinal tract, and an increase in their population in the human gastrointestinal tract is associated with a decline in the frequency of enteric infections and an improvement in intestinal health. Only about 15 of the 50 identified Bifidobacterium species and subspecies are often utilized commercially as probiotics (Sharma, Wasan, & Sharma, 2021). These include Bifidobacterium breve, Bifidobacterium longum, Bifidobacterium animalis, Bifidobacterium bifidum, and Bifidobacterium adolescentis. The coryneform group of non-motile, non-sporing, non-acid-fast, irregularly shaped organisms is represented by the genus Brevibacterium. Brevibacterium spp. Gram-positive rods are like other coryneform bacteria, but their staining pattern and shape can change depending on the age and health of the cells. They are high G+C organisms. Brevibacterium are strictly aerobic, mesophilic, catalase-positive microorganisms with an ideal growth range of 20 to 35 °C. The majority of species can thrive in a wide pH range and are salt-tolerant (> 10%). There are 35 identified species. Several of them have been categorized as pathogens and are important medically. However, one species, Brevibacterium linens, is crucial for the production of cheeses like Limburger and Munster, making it important in fermented foods. B. linens create a yellow, orange or crimson pigment on the surface of the cheese in these products, giving the cheeses their distinctive color. Their capacity to hydrolyze proteins and metabolize amino acids, particularly those containing sulfur, has a role in the flavor and ripening of a variety of cheeses.

4.8 BACTERIA

Gram-positive, catalase-positive, non-motile, non-spore-forming, aerobic cocci make up organisms like Kocuria, Micrococcus, and Staphylococcus. Despite these similarities, Kocuria, Micrococcus, and Staphylococcus are all members of the Actinobacteria, while Kocuria is a Firmicute (Chun et al., 2015). They are all coagulase-negative, which sets them apart from coagulase-positive cocci which are linked to the formation of enterotoxins. A small number of species are important for fermented foods. Their involvement is primarily restricted to the production of dry fermented sausages, though some may accidentally contaminate cheese and other foods. Since they are not fermentative, these bacteria are not used to produce acids in fermented meats; instead, they are used to improve flavor and color development. They specifically make aromatic molecules, hydrolyze proteins and lipids, and convert nitrate to nitrite. Because they are eukaryotes, fungi are fundamentally different from prokaryotes. In particular, fungi reproduce differently from bacteria due to their complex composition. Molds and yeasts are both fungi, however, yeast is unicellular and non-filamentous, while molds are multicellular and filamentous. Yeast exclusively produces sexual spores or is non-sporulating, but molds can generate both asexual and sexual spores. Conidia are the common name for the asexually produced mold spores. These are the spores that Penicillium and Aspergillus create. Conidiophores, where conidia are born, can be recognized as belonging to a particular mold based on their characteristics. Sporangiospores are produced in sporangia by Rhizopus and Mucor(Dennis & Blijham, 1980). In addition to ascospore production, yeasts can engage in asexual reproduction known as budding, a process in which a portion of the cell cytoplasm protrudes from it before eventually separating as a new cell. Budding is a trait of the species and can either be multilateral or polar.

4.9 SACCHAROMYCES

Yeasts from the genus Saccharomyces are among the most crucial microbes present in fermented foods. The manufacture of beer, wine, and spirits depends on yeasts. Saccharomyces cerevisiae, the dominant species, is also frequently employed as a model organism in biology and genetics, and its physiological and biochemical characteristics, as well as its genome sequence, have been investigated. Saccharomyces and other yeasts are significantly more complex than bacteria at the gene level. They can be diploidal or polyploidal and have 16 chromosomes. As a result, they have genomes that are several times bigger than those of bacteria, with more than 6000 protein-coding genes as opposed to around 2000 in the majority of lactic acid bacteria. Since more than a century ago, modifications to Saccharomyces taxonomy and nomenclature have been frequent. This was particularly true when yeast taxonomy was based mostly on morphology and physiology before the development of molecular typing techniques. Although S. cerevisiae (Figure 4.5) has long been a key species in applications for manufacturing wine, beer, and bread, the individual strains utilized to make these products are very distinct from one another and from laboratory strains of S. cerevisiae. It has become even more challenging to remember the specific species connected to a specific fermentation because of the frequent changes in nomenclature.

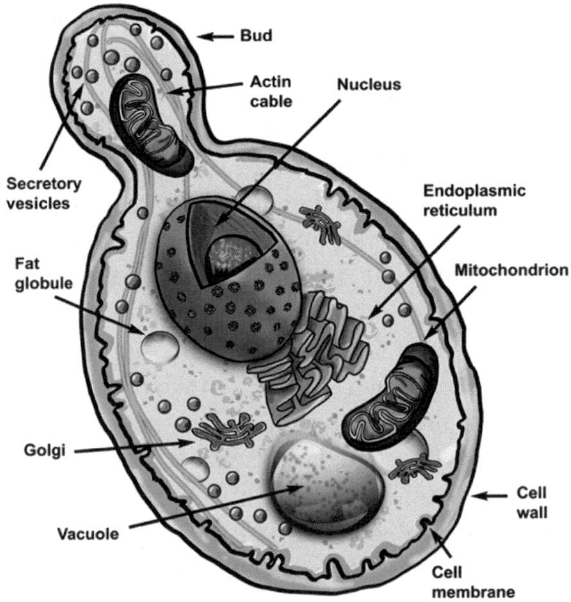

Figure 4.5. S. cerevisiae.

(Source: *Walker GM, Stewart GG. Saccharomyces cerevisiae in the Production of Fermented Beverages. Beverages. 2016; 2(4):30. https://doi.org/10.3390/beverages2040030, Creative Commons License)*

Saccharomyces ellipsoideushad species' status before the 1970s was acknowledged as one of the main wine yeasts. The previous term is still in use, but it was later reclassified as S. cerevisiae. The classification of a different group of wine strains previously known as Saccharomyces uvarum was changed to Saccharomyces bayanus (Magyar & Toth, 2011).

Saccharomyces carlsbergensis, also known as Saccharomyces uvarum, was considered to be the species utilized in the fermentation of lager. The two species then combined to become S. uvarum before being once more categorized as Saccharomyces pastorianus and then Saccharomyces bayanus. The latter is currently thought to be a valid species name for the lager yeast. These yeasts are typically spherical or ovoid, however, they can occasionally become elongated with pseudohyphae.

The modes of sugar fermentation and the uptake of carbon sources are crucial elements in speciation. Ecologically speaking, foods frequently include Saccharomyces and other yeasts. In some fermented foods, like wine, their normal presence on the grapes and other machinery is enough to start the fermentation process. Wine fermentations that occur naturally or spontaneously may involve a variety of yeasts, usually successively, with one species predominating for a while before giving way to others. Starter culture yeasts are employed in the majority of contemporary wine fermentations, as well as in beer and bread fermentations, and these yeasts are based on their physiological and biochemical characteristics. Other fermented foods, such as fermented meats, soy-fermented meals, and a wide variety of cheeses, also benefit from yeast taste and quality contributions. Yeasts play a significant role in the deterioration of both fermented and non-fermented foods. Rhodotorula, the yeast that produces pink pigment, is responsible for one of the main problems in sauerkraut. Even while Aspergillus species play a less significant part in the production of fermented foods, they are still a component of some of the fermented foods that are consumed the most around the world. Aspergillus oryzae and Aspergillus sojae, two principal species, play a vital role in the manufacture of various fermented foods common to Asian cuisines, including soy sauce, miso, and rice wines. The production of koji is the first step in the fermentation of these foods, which are ingested by billions of people. More than 330 different species of Aspergillus and more than 350 different species of Penicillium exist. Similar to yeasts, speciation is based on spore type, appearance, and structure. However, the biochemical and physiological characteristics of the fungi are typically less diagnostic. Recent research in next-generation sequencing technology and the use of bioinformatics tools have made it possible to better comprehend the dynamic nature of food fermentations(Kim et al., 2014).

Pasteur, Buchner, Schwann, and other microbiologists and biochemists were the first to study the biochemistry of fermentation and the physiology of microorganisms. The fermentation of sugars by microbes was necessary for the production of fermented milk products, beer, and wine, which were industrial and cultural products. Therefore, identifying the organisms and pathways was crucial to understanding how to regulate or improve the production of these products as well as to prevent or postpone spoilage. Lactic acid bacteria are obligate species that do not get their energy from oxidative or respiratory activities. Technically, these species obtain ATP through other metabolic processes, such as the precursor-product exchange systems. These pathways were developed by bacteria to conserve the energy units that would typically be needed to carry out metabolic tasks. However, these cells primarily produce ATP through processes at the substrate level that take place during fermentation. Hexoses are digested by homofermentative lactic acid bacteria using the enzymes of the glycolytic Embden-Meyerhof-

Parnas (EMP) pathway. Aldolase cleaves fructose 1,6 diphosphate into two triose phosphates, which later act as substrates for ATP-producing activities. The EMP route produces two moles of pyruvate and two moles of ATP. The enzyme lactate dehydrogenase then reduces the pyruvate to L or D-lactate. During homofermentative metabolism, more than 90% of the substrate is transformed into lactic acid.

There are three basic modes of sugar transfer in lactic acid bacteria. The majority of mesophilic, homofermentative lactic acid bacteria, including lactococci and pediococci, use the phosphoenolpyruvate (PEP)-dependent phosphotransferase system (PTS) to transport lactose and glucose. Other lactic acid bacteria, however, use symport-type or ATP-dependent mechanisms to transport sugars. The PTS is made up of a series of cytoplasmic and membrane-associated proteins, and its function is to transfer a high-energy phosphate group from PEP, the initial donor, to the incoming sugar in a sequential manner. A sugar-phosphate builds up inside the cell as a result of the phosphorylation stage. The PTS components that are connected with the membrane are substrate-specific. The Enzyme II complex, which may be made up of distinct proteins or domains, is formed by these proteins. One of the most crucial microbiological activities is the transfer of nutrients from the external medium through the cytoplasmic membrane and into the cytoplasm. Nearly 20% of the genomes of many microorganisms, including lactic acid bacteria, are devoted to transport-related tasks. Cell membranes contain 100 or more transport permeases, translocators, and auxiliary proteins. Transport is hard since it typically requires energy (ATP) to move molecules from one side of a membrane to the other. This is because nutrient transport often takes place against a gradient in concentration, which means that the concentration of the substrate is significantly lower in the outside medium than it is inside. Even while there are rare instances of passive or facilitated diffusion, when the concentration gradient drives transport, it is usually necessary to engage in an active, energy-intensive process.

Given their limited and ineffective means of producing energy, lactic acid bacteria have a very high metabolic cost of transport. The amount of ATP produced by glycolysis is just two molecules for every molecule of digested glucose or hexose (Tappy & Rosset, 2017). The net energy gain by the cell may be reduced by as much as 50% if ATP or its equivalent is needed to transport mono and disaccharides across the cytoplasmic membrane. If the energy source that powers solute transport is independent, the cell can store the extra energy. Many lactic acid bacteria produce several metabolic products that must be expelled from the cell as a result of the metabolism of substrate or precursor molecules. These compounds are sometimes not further metabolized; in other situations, they might even be harmful if allowed to build up inside cells. In any case, a significant concentration gradient is created, with the product concentration within being significantly higher than the concentration outside. Therefore, the concentration gradient acts as a downhill force to push the efflux of product molecules. However, in these conditions, the carrier for efflux possesses an affinity for both the substrate and the product. It works in the other direction; thus, product efflux can cause the precursor substrate to be taken against the gradient. The transport machinery plays a crucial role in the regulation of sugar metabolism in lactic acid bacteria. Lactose fermentation in L. lactis is controlled biochemically and genetically at the transcriptional level by a signal transduction cascade that is mediated by the lactose transport system (Kleerebezemab et al., 2000).

4.10 ENZYMES

Several glycolytic enzymes have allosteric properties, meaning that their actions depend on the intracellular concentration of specific glycolytic products. In theory, the pathway is controlled so that when there are plenty of substrates, glycolysis is completely activated, and when there aren't enough, glycolysis is reduced. Fructose 1,6-diphosphate or FDP will therefore be formed in large quantities inside of cells during milk fermentations where there is an abundance of lactose. Pyruvate kinase and lactate dehydrogenase are both activated by high amounts of FDP. If there is enough substrate, activation of these enzymes keeps glycolysis active. Pyruvate kinase activity reduces in the absence of FDP and the presence of phosphate, generating a metabolic checkpoint in glycolysis where PEP would typically be used to drive ATP synthesis. Glycolysis is stopped, and PEP ultimately builds up as a result. The PTS transport proteins, and in particular the cytoplasmic component, HPr, are involved in a metabolic pathway that directly controls lactose metabolism. HPr is phosphorylated by Enzyme I to begin the PTS cascade. Fructose 1,6 diphosphate, a glycolytic precursor that only reaches high concentrations when the substrate is abundant, activates HPr kinase. Repressor proteins can directly control the genetic regulation of sugar metabolism in lactic acid bacteria. For instance, the repressor protein LacR, which is produced by the lacR gene, negatively regulates the lactose PTS operon in L. lactis. Lactose causes the expression of lacR to be inhibited and the transcription of the lac operon to be stimulated. However, LacR is expressed and lac gene transcription is suppressed when lactose is missing or when cells are grown on glucose. Although a CRE site can also be found close to the lac operon's transcriptional start site, indicating that the lac genes are prone to CcpA-mediated repression, it appears that LacR may be primarily controlling sugar metabolism (Andersen et al., 2011).

The rate-limiting stage in S. cerevisiae glycolysis is glucose transport. S. cerevisiae contains up to 18 active glucose transporters. Because they transport additional hexoses, the transporters are known as hexose transporters or HXT proteins. These hexose transport systems use assisted diffusion to carry glucose. These systems are solely driven by the concentration gradient and are energy-independent. Some of these transporters are constitutive or inducible, some have low or high substrate affinities, and some serve as molecular sensors rather than as transporters. The initial concentration of fermentable carbohydrates is high during beer or wine fermentations, but it gradually drops as the fermentation goes on. For Saccharomyces to ferment all of the substrates, it needs transport systems that can operate at high, low, and intermediate substrate concentrations. The regulatory characteristics of sugar metabolism in S. cerevisiae are very distinct from those of other organisms in terms of the actual processes and the degree to which other catabolic pathways are impacted. In general, catabolite suppression happens when there are two or more fermentable sugars in the medium and the cell has to choose which sugars to ferment first. Since glucose is typically the favored sugar, catabolite suppression affects the genes that code for the metabolism of other sugars. Catabolite suppression in lactic acid bacteria is mediated by the phosphorylation state of HPr working in tandem with CcpA. Catabolite repression, also known as glucose repression, has a broader purpose in S. cerevisiae. In addition to suppressing the transcription of genes involved in sugar metabolism, glucose also inhibits the activity of genes encoding for several other critical processes, such as mitochondrial oxidation, the citric acid cycle, the glyoxylate cycle, and glycerol consumption.

The majority of fermented foods and beverages are made via sugar metabolism, specifically lactic fermentation and ethanolic fermentation. Other metabolic processes, however, are also crucial. Protein metabolism not only meets cellular needs but also contributes to cheese flavor and texture. Lactic acid bacteria execute fermentations in other dairy products like buttermilk and sour cream that produce diacetyl, a substance that gives these foods their distinctive flavor. The fermentation of propionic acid in Swiss-style cheeses produces many metabolic by-products that also contribute to the distinctive characteristics of these cheeses. Fungi produce a variety of compounds with aromatic and flavor qualities even though they do not ferment substrates. Many species of lactic acid bacteria have reverse reactions in addition to hydrolyzing and metabolizing carbohydrates. Species that can produce exopolysaccharides are found in almost all genera involved in food fermentations (Tieking & Ganzle, 2005). While many bacteria may survive on ammonia or other inorganic nitrogen sources, lactic acid bacteria cannot assimilate inorganic nitrogen and must rely on preformed amino acids to meet their amino acid needs. This means that lactic acid bacteria must first be able to digest proteins and peptides and then also be able to transfer the free amino acids and tiny peptides generated during proteolysis, as most foods only contain a small pool of free amino acids. The proteolytic system is made up of the enzymes and transport mechanisms required for these tasks. Casein, a milk protein, is the main substrate for these bacteria.

CHAPTER

5 Fermented Products with Plants

Contents

5.1 INTRODUCTION

Different civilizations have a long history of having access to fermented foods. Because of cultural customs, different foods are more or less acceptable. A product that is liked by consumers in one culture could not be liked by consumers in another one. There is an endless amount of fermented food products. Due to factors like food categories, final product form and qualities, type of ingredients utilized, and cultural variety, the manufacturing techniques for fermented products vary greatly. The production of fermented foods involves a variety of manufacturing techniques. Baked yeast bread is a dietary supplement for the population in the regions that produce wheat (Linko et al., 1997). Various baked bread varieties include flat bread, specialty bread, and standard yeast bread. To satisfy consumers' predilection for cuisine that has the appearance of being home-cooked, even chilled or frozen doughs are now readily accessible. Other types of bread, such as steamed bread and boiled bread, are available in regions with reduced energy availability. Fried bread is typically eaten for breakfast or as a snack. Today, many varieties of wheat are available to fit production conditions in various regions with distinct climatic conditions as a result of centuries of breeding selection. To satisfy product requirements, wheat used to make bread might be either hard, soft, or a mix of the two.

Figure 5.1. Wheat flour.

*(**Source:** Iuga M, Boestean O, Ghendov-Mosanu A, Mironeasa S. Impact of Dairy Ingredients on Wheat Flour Dough Rheology and Bread Properties. Foods. 2020; 9(6):828. https://doi.org/10.3390/foods9060828, Creative Commons License)*

5.2 WHEAT FLOUR

Wheat flour (Figure 5.1) is made from wheat kernels after the bran has been removed (Van Der et al., 2005). This flour is the main component in bread baking. Whole wheat flour is currently the flour of choice for producing bread for some health-conscious consumers. To improve the fiber level, wheat bran is also added. The main components used in the production of various wheat-based bread and associated foods are wheat flour, yeast, salt, and water. The other ingredients differ depending on the product type. These could be categorized as additions, optional components, or processing aids. Every country has its laws and specifications. The term flat bread refers generally to bread products that do not rise as much as regular bread. In the Middle East and other places with limited access to energy, flat bread is a staple food. The term specialty bread refers to flat bread in developed nations. The dough is created similarly to how bread is normally made. However, before being roasted directly in the fire or an oven, the dough is flattened and occasionally layered. Danish pastries and croissants are examples of items that come from modifying the fundamental bread-making procedure. Although the processes for making the dough are similar, the components are not. In the Chinese community, steamed bread is common. The main food staple in China's northern provinces is plain steamed bread (He et al., 2003). Steamed buns with filling, however, are a specific food in some regions of China. The process of dough solidification is the key area where steamed bread production differs from conventional bread production. While baking is used to make regular bread, steaming is used to make steamed bread. Because the temperature utilized is not high enough to trigger the browning reaction, there is no brown crust on the surface of steamed bread. Because the bread is soft at this temperature, steamed bread is always consumed hot or kept in a steamer. Steamed bread may occasionally be deep-fried before consumption. When steamed bread cools, it hardens and becomes less tasty. Steamed bread can be made using a variety of techniques.

Depending on the proportion of ingredients like wheat and soybeans, the fermentation and extraction processes, and the flavoring ingredients like caramel and others are employed, there are several types of soy sauce. However, the production processes are comparable. To hydrolyze the soy proteins and starches into smaller flavor components, boiled soybeans are fermented in salt or brine under controlled conditions to produce soy sauce. The fermented soybeans are subsequently used to extract the soy sauce, which is then standardized and packaged. A classic Japanese product made from fermented soybeans is natto(Taniguchi et al., 2019). Soybeans are characterized by a mucilaginous, sticky coating that is now thought to provide health advantages. It has a short shelf life and is made by quickly fermenting cooked soybeans with Bacillus natto. In Japan (Hamamatsu region), entire soybeans that have been fermented are used to make hama-natto. Tou-chi or dou-chi is a very similar product in Chinese culture. To make it, cooked soybeans are fermented in salt, brine, or soy sauce before being dried as individual beans. In contrast to dou-chi, which may or may not have ginger flavoring, hama-natto contains ginger in the flavoring.

Figure 5.2. Fermented soy pastes.

(Source: *https://upload.wikimedia.org/wikipedia/commons/e/ee/Doenjang-withbeans.jpg, Creative Commons License)*

5.3 SOY

Fermented soy pastes (Figure 5.2) are common in both Chinese and Japanese cultures, and they are both manufactured similarly. The Japanese utilize their fermented soy paste, miso, to flavor and marinate fish, and to a lesser extent, to make miso soup. The miso soup is an essential food of traditional Japanese cuisine. Dou-pan-chiang, a fermented soy paste, is mostly used as a condiment in Chinese cooking. Wing beans can also be used to make dou-pan-chiang. Tofu, which is manufactured by coagulating the soy protein in soymilk with calcium or magnesium sulfate, is fermented to make sufu. Its fermentation procedure is comparable to that of feta cheese. Both goods are aged in sealed vessels of brine. Sufu can be packaged with seasoning additives. A classic Chinese dish called tofu is made by quickly fermenting tofu in brine created by fermenting vegetables under specific conditions (Zhao et al., 2014). During this brief fermentation, the tofu undergoes a small amount of hydrolysis and creates the flavor chemicals that make it distinctive. These molecules volatilize during the deep-frying process of this raw tofu, giving it its distinctive smell. It is typically eaten with soy and chili sauces. People in Indonesia frequently eat tempe, a traditional dish. It is produced by fermenting cooked soybeans that have been covered in plastic or dried banana leaves. The mycelia that the moldRhizopus oligosporus develops pierce the block of soybeans (Handoyo & Morita, 2006). The block is encircled by mold mycelia as well. This is comparable to how molded cheese ferments. Vegetarians in the West are gradually coming to recognize tempe as a diet. It is typically eaten as a deep-fried food item.

In the past, several societies developed fermented vegetables to preserve vegetables when they weren't available or due to climate restrictions. Some of these items were once traditional foods in their original civilizations but are now commonly used in other cultures. These fermented vegetables are made with salt. The main microbial activity in these processes is natural lactic acid fermentation, which produces enough lactic acid to reduce the pH. These two substances combine to make lactic acid, which creates an environment that can prevent the growth of other microbes. As the raw ingredients, leafy vegetables, fruits, and roots are employed. Occasionally, starter cultures are employed. Some vegetable items are preserved with vinegar, while others utilize chili powder and other spices. Sauerkraut is a traditional fermented vegetable product from Germany that has become popular in other cultures and can be consumed with other dishes. Each lactic acid bacterium takes control of the fermentation until the result is inhibitory to its growth and establishes a favorable environment for another lactic acid bacterium to take control. Until the majority of the available fermentable sugars are used up, fermentation will continue. Sauerkraut production is not risk-free, so sanitation steps must be followed to prevent spoiling (Penas et al., 2017).

5.4 PICKLES

Western-style pickles are made by first salting the cucumbers to be pickled in salt stocks for long-term storage, desalting, and then bottling in sugar, vinegar, and optionally spices. Lactic acid fermentation continues to be the method of fermentation. It is more prone to deterioration nevertheless since the cucumber wax coating may trap air inside. Spoilage can happen when cucumbers are salted-cured, thus steps should be made to prevent it from happening. Western-style pickles are typically mildly heated to sterilize or pasteurize the food due to the high acidity, low pH, and high salt content of the product. Many pickled vegetables are produced by the Chinese. Vegetables of many varieties are used as basic materials. Depending on the final product, the fermentation can either be a dry-salting or a brining procedure. However, lactic acid fermentation continues to be the method of fermentation. Desalting is typically not used in the production of Chinese-style pickled vegetables, which is the main distinction between those products and Western-style pickles. Some vegetables prepared in the Chinese method are converted into intermediate moisture products, which is not done in the Western method. One of the oldest techniques for preparing and preserving food still in use today is fermentation. Fermentation is a process that turns raw food resources like milk into cheese and yogurt, grape juice into wine, wheat into bread, barley/malt into beer, and soybeans into soy sauce from unstable, occasionally non-functional forms. The three main functions of food fermentation are preservation, enhancement of sensory qualities i.e., flavor, texture, or color, and transformation of nutrients via improvement of digestibility and biofortification (Tolve et al., 2021).

5.5 SENSORY QUALITIES

One sensory quality that is heavily altered by fermentation is flavor. Substrates like carbohydrates, proteins, lipids, organic acids, amino acids, phenolic compounds, glycosides, etc. are converted into volatile and non-volatile flavor chemicals that affect food acceptance by influencing not only taste but also odor. These flavoring substances give sensory qualities, including sweetness (mannitol), sourness (lactic acid), umami (l-glutamic acid), bitterness (hydrophobic peptides), fruitiness (esters), and sulfurous odor (volatile sulfur compounds). The concentrations of the taste chemicals and the food matrices in which they are formed determine whether they are desirable or undesirable. Food acceptability and selection by consumers are heavily influenced by flavor i.e., aroma and taste. In food manufacturing, uniformity of flavor is crucial to food quality. A growing trend is the diversification of flavors due to customer demand for new and interesting flavors. To guarantee flavor consistency and to provide flavor diversity, it is necessary to have a deeper grasp of how flavors are formed during fermentation. The production of fermented foods might rely on the microflora that is naturally present on or in the raw materials, such as raw milk cheeses.

Back slopping, or the transfer of the natural microflora from the fermenting materials to the fresh starting materials is a typical procedure in domestic tasks or the manufacture of several traditional fermented foods such as bread and fermented milk items (Tou et al., 2007). To assure uniformity in quality and better control, raw materials are now inoculated with starter cultures during the industrial manufacturing of fermented foods.

5.6 FOOD FERMENTATION

Food fermentation microorganisms include bacteria, yeasts, and fungi. There are four basic types of food fermentation: bacterial fermentation, yeast fermentation, fungal fermentation, and mixed culture fermentation. A variety of fermented foods, including cheese, fermented milk, fermented vegetables, natto, and fermented sausages, are produced as a result of bacterial fermentation. Yeast fermentation frequently converts raw ingredients rich in sugar into alcoholic beverages or flour into bread. The majority of fermented foods, particularly those that are traditional or native, are made utilizing microbial cultures that include bacteria, yeasts, and fungi. Examples of fermented foods made by combining different cultures include kefir and koumiss, Swiss cheese, soy sauce, wine, and bread. Based on the primary products formed or pH changes, the process of food fermentation can also be divided into alcoholic, acidic, and alkaline fermentations. Alcoholic beverages fall under the category of alcoholic fermentation carried out by yeasts, with ethanol serving as the primary product of sugar metabolism. The process of lactic fermentation, which uses lactic acid bacteria, falls under acid fermentation, in which lactic acid is the primary by-product of the conversion of sugar. For example, during the fermentation of several fermented soybean products such as tempeh, natto, fermented tofu/fermented soybean curd, ammonia is often produced due to amino acid deamination. This is distinct from the lactic acid catabolism by yeasts that cause the pH to rise in ripened cheeses. Foods can be fermented in a liquid condition, a solid-state, or a mixed state. Fermented milk, yogurts, and alcoholic beverages like beer and wine are produced via liquid state food fermentations. Bread, fermentation of cocoa and coffee beans, and some soybean products are examples of solid-state food fermentations. Some food fermentation processes involve phase transitions from liquid to solid or from solid to liquid states.

The microorganisms that cause food to ferment and flavor formation in the form of pure, defined, or undefined mixed cultures are bacteria, yeasts, and fungi. Food substrates, as well as intrinsic and extrinsic elements like pH, temperature, and water activity, affect the number of microorganisms that take part in food fermentation. Bacteria produce acids, engage in proteolytic and lipolytic processes, and modify amino acids, all of which contribute to the flavor formation in fermented foods. Alcoholic drinks like beer, wine, and sake are all fermented by yeasts, as are bread, soy sauce, ripened cheeses, and other foods. Non-Saccharomyces yeasts, however, also contribute to the fermentation of some foods, including bread, kefir and koumiss, soy sauce, fermented sausages, and ripened cheeses (Yin et al., 2020). The dairy yeasts, such as Candida kefyr, Debaryomyceshansenii, Geotrichumcandidum, Kluyveromyces lactis and K. marxianus, and Yarrowialipolytica, as well as yeasts like Candida milleri and soy sauce yeasts like Candida versatilis and Zygosaccharomyces rouxii, are among this non-Saccharomyces yeast. These yeasts alter sugars, amino acids, glycosides, and phenolic acids during fermentation to produce alcohols, esters, carboxylic acids, lactones, aldehydes, volatile sulfur compounds, and volatile phenols, which contribute to flavor. The microflora that supports the fermentation of many types of fermented foods also includes fungi. Penicillium roqueforti and Penicillium camembert are used to create mold-ripened cheeses, such as Camembert and Brie. P. camembert, Penicillium nalgiovense, and Penicillium chrysogenum are used to ferment a variety of sausages. Aspergillus oryzae, Aspergillus sojae, Mucor sp. and Actinomucor sp., Monascus sp., and Rhizopus oligosporus are among the fungi used to produce fermented soy products. Aspergillus oryzae produces soy sauce, miso, soybean paste, and fermented soybeans. Alcoholic beverages made from grains, like rice wine, always contain A. oryzae (sake). During the solid-state phase of fermentation, these fungi produce enzymes that degrade substrates like proteins, lipids, and polysaccharides in the raw materials.

In addition to microbial fermentation, raw ingredients and production methods (Maillard reaction) can contribute to the flavor of fermented foods because of chemical and biological modifications (Tuohy et al., 2006). A wide variety of foods, including fermented milk, yogurt, cheese, and conventional lactic butter, are included in the category of fermented dairy products. Different types of fermented dairy products have different flavors. Traditional lactic butter aroma is caused by diacetyl, while the flavor of plain yogurt is primarily influenced by both lactic acid and acetaldehyde. Depending on the type of cheese, cheese flavors can be complex and varied. The peptides produced by protein breakdown can produce bitterness. On the other hand, the fundamental flavors of cheeses, particularly glutamate, are influenced by the amino acids from peptidolysis and proteolysis. The lipolytic odor is caused by short-chain fatty acids from lipolysis, such as butyric acid. While the aroma of ripened cheese is produced by volatile sulfur compounds, blue cheese flavor is characterized by substantial quantities of methyl ketones. Esters give Italian-style cheeses their characteristic texture, whilst propionic and acetic acids are responsible for the distinctive flavor of Swiss cheeses. Products made from fermented meat mostly include fermented ham and sausage. The flavor of these foods is influenced by lactic acid, fatty acids, branched-chain aldehydes and acids, ketones, and amino acids. In Southeast Asia, fermented fish products like fish sauce are popular. Halophilic lactic acid bacteria, aldehydes, ammonia, hydrocarbons, and sulfur compounds contribute to the flavor of fish sauce.

Soy sauce, Japanese miso, soybean paste, Japanese natto and Korean cheonggukjang, fermented black beans, and fermented soybean curd or tofu are a few examples of the variety of products made from fermented soybeans. Branched-chain fatty acids, such as 2- and 3-methylbutanoic acids, are thought to be the source of the natto odor. Alcohols, 4-hydroxy furanone, methionol, and volatile phenolic chemicals, when combined give soy sauce its sweet and spicy flavor are the major ingredients that give it its umami flavor. Products made from fermented soybeans contain volatile sulfur compounds. The flavor of fermented alcoholic beverages like beer, wine, and grain-based spirits is enhanced by a variety of fermentation-derived volatiles, including alcohols, carboxylic acids, carbonyls, esters, and even volatile sulfur compounds. The volatile ingredients that add flavor are esters generated by yeast. Branched-chain esters, such as isoamyl acetate, ethyl esters of fatty acids like ethyl octanoate and 2-phenylethyl acetate are examples of typical esters. In addition to the acetic acid, ethanol, and succinic acid produced by the lactic acid bacterial fermentation, lactic acid is what gives fermented foods like sauerkraut and kimchi their sour flavor. Fermentation results in the production of alcohols, aldehydes, and esters, which are the principal flavoring agents in bread. These volatiles from yeast give bread its sweet, and malty flavor. Acetic acid is widely acknowledged as a vinegar flavoring element. However, acetic acid alone does not entirely account for vinegar flavor. The content of volatile components varies between vinegar types. Volatile substances formed during vinegar fermentation, such as aldehydes, carboxylic acids, and esters, as well as volatile substances carried over from raw ethanolic materials like cider, wine, rice wine, and malt mash, have an impact on the flavor of vinegar.

5.7 METABOLISM

Microorganisms use carbohydrates as fuel for growth. As a result, a variety of volatile and non-volatile compounds, including acids, alcohols, and esters, can be formed. Based on the microorganisms involved, different flavor compounds are produced in different concentrations. Hexoses are utilized by lactic acid bacteria via either the homofermentative or heterofermentative pathways, with the former producing lactic acid as the main end product and the latter producing lactic acid, acetic acid, ethanol, and carbon dioxide (Zuniga et al., 1993). Pentoses are converted to lactic and acetic acids by some lactic acid bacteria. In addition to creating lactic acid, acetic acid, and carbon dioxide, heterofermentative lactic acid bacteria can use fructose as an electron acceptor and convert it to mannitol. A sugar alcohol known as mannitol has a variety of uses, including texturizing and calorie-reduced sweetening. In fruit and vegetable fermentations where fructose and sucrose are readily available, mannitol can be formed. If enough mannitol is produced, it may contribute to sweetness. On the other hand, facultative Lactobacillus plantarum can convert mannitol to lactic acid. Lactic acid bacteria convert carbohydrates and lactic acid into propionic acid, CH_3COOH, and CO, which are essential ingredients for the flavor and texture of Swiss-style cheese. Lactic and acetic acids are produced when carbohydrates are fermented by bifidobacteria; the latter is what gives fermented milk manufactured with bifidobacteria their vinegary flavor. Zymomonas bacteria may turn glucose into ethanol or a mixture of glucose and fructose into ethanol and sorbitol, a sweetener. Beer and cider spoiling is a result of the Zymomonas ability to produce excessive amounts of acetaldehyde during the fermentation of sugar.

Yeasts are able to produce a wide range of non-volatile and volatile flavor compounds from sugars, including glycerol, higher alcohols (3-methylbutanol), carboxylic acids (octanoic acid), ethyl esters of fatty acids (ethyl octanoate), and carbonyls (acetaldehyde). Two stages of sugar fermentation are required to generate vinegar: first, yeasts ferment sugar-rich substrates to produce ethanol, and then acetic acid bacteria convert ethanol to acetic acid. Proteolysis and peptidolysis linked to microbial transformation produce peptides and amino acids that give fermented, protein-rich foods like cheeses and fermented soy products their fundamental flavors. Hydrophobic peptides with high concentrations of hydrophobic amino acids like proline cause bitterness, notably in cheeses, while glutamate-rich peptides provide a peculiar taste. The primary flavors of amino acids are umami (glutamic acid), sourness (aspartic acid), sweetness (alanine), bitterness (arginine), and sweetness (alanine). The primary flavor component in cheeses and fermented soy products is protein-derived glutamic acid. Many aromatic compounds are precursors to amino acids, which are generated by bacteria, yeasts, and fungi through a variety of processes or found in free form in dietary items. The specific aromatic compounds, such as acids, alcohols, aldehydes, and esters, that are formed depend on the amino acid precursors and the microorganisms.

5.8 AROMA

The primary aroma precursors for branched-chain alcohols and aromatic alcohols produced by yeasts using the Ehrlich pathway are branched-chain amino acids i.e., leucine, isoleucine, and valine and aromatic amino acids like phenylalanine, tryptophan, and tyrosine (Landaud et al., 2008). Similar amino acid catabolism is possible in lactic acid bacteria, although this usually necessitates the availability of an amino acceptor, such as -ketoglutaric acid. These amino acids can be used by microorganisms to create aldehydes, carboxylic acids, and esters in addition to branched-chain and aromatic alcohols. Other amino acids can also be broken down to produce different flavor-enhancing compounds, such as threonine being broken down into acetaldehyde and glycine by yogurt cultures, serine being degraded to acetoin by L. plantarum, and aspartic acid being converted to diacetyl and acetoin by Lactococcus lactis. Tetragenococcushalophilus, an interesting soy sauce lactic acid bacterium, has the ability to decarboxylate aspartic acid, which has a sour flavor, into alanine, which has a sweet flavor. It's important to note that propionibacteria employed in the fermentation of Swiss cheese produce aromatic molecules like carboxylic acids from branched-chain and aromatic amino acids similarly to how lactic acid bacteria do through transamination in the presence of an amino acceptor like α-ketoglutaric acid. Aspartate is additionally transformed by protonibacteria into succinate, acetate, ammonia, and carbon dioxide. When an amino acceptor like α-ketoglutaric acid is present, the sausage starting cultures Staphylococcus carnosus and Staphylococcus xylosus may also catalyze the transamination of branched-chain amino acids into aldehydes and acids.

Many of these sulfur compounds, including methionol, methional, thioesters, thiols, and sulfides, have direct or indirect microbial origins. The primary precursor to volatile sulfur compounds is methionine. Although demethiolation of methionine is an alternate process that results in the formation of methanethiol, yeasts mostly make methionol from methionine by the Ehrlich pathway. On the other hand, the primary mechanism in lactic acid bacteria and brevibacteria appears to be methionine demethiolation (elimination by lyases). Both Saccharomyces and non-Saccharomyces yeasts can make methylional and methionine-derived esters (ethyl 3-methylthio-1-propanoate and 3-methylthio-1-propyl acetate). Yeasts release non-volatile, cysteine-bound conjugate volatile thiols (Swiegers et al., 2006). Two typical organic acids i.e., malic and citric acid are vulnerable to microbial attack during fermentation, which might alter their sensory qualities. Yeasts, particularly Schizosaccharomyces pombe and Schizosaccharomycesmalidevorans, have the ability to convert malic acid to ethanol and carbon dioxide (maloethanolic fermentation), however, S. cerevisiae yeasts have a limited capability to do so. On the other hand, lactic acid bacteria, particularly Oenococcusoeni, can break down malic acid to create lactic acid and carbon dioxide (malolactic fermentation). Highly acidic wines made in cold temperature zones benefit from the conversion of malic acid to less acidic lactic acid or ethanol because it lessens the bitterness and enhances the mouth feel. Aside from lactic acid, citric acid is a precursor of a number of taste chemicals, such as acetic acid, diacetyl, acetoin, and succinic acid. By lactic acid bacteria, such as L. mesenteroides subsp. cremoris and Lactococcus lactis subsp. lactis var. diacetyllactis in dairy fermentation, O. oeni in wine fermentation, and spoilage lactic acid bacteria in beer fermentation, this organic acid is normally broken down. Although it is a component of the aromatic profile of other fermented foods like cheese, fermented vegetables, and especially bread, acetic acid is not needed in alcoholic beverages because it imparts a vinegary off-flavor. Butter, cheese, and yogurt contain an aroma molecule called dialdehyde. However, because of its extremely low detection threshold, diacetyl is particularly unfavorable in beer and contributes to an off-odor.

Although it has a buttery smell, acetoin has a higher detection threshold than diacetyl, therefore in most cases, its concentration is not high enough to give fermented foods and beverages a flavor. Some lactobacilli are capable of converting citric acid to succinic acid, a flavor enhancer that imparts a brothy flavor. Some plant-based matrices, like fruit juices, contain aroma components in free or bound, non-volatile, and odorless forms (glycosides or glycoconjugates). The oxygenated terpenes, C13-norisoprenoids, benzene derivatives (volatile phenolics), and aliphatic alcohols are frequently found as the aglycones of the glycosides. Due to chemical and enzymatic degradation, these glycosides can produce aromatic molecules during fermentation and are used as aroma precursors. During alcoholic and malolactic fermentations, glycosidases from both S. cerevisiae yeasts and the lactic acid bacteria O. oeni can hydrolyzeglycosidically bound aroma molecules like linalool, citronellol, and hexanol. Alcoholic beverages made from fruit are enhanced by the highly oxygenated terpene contents. Numerous variables, including yeast strains, malolactic bacterial strains, and fermentation conditions (pH, temperature, and ethanol), affect the liberation of aroma compounds on the hydrolysis of sugar-bound aromatic precursors.

The primary components of phenolic acids, which are often attached to plant cell walls, are ferulic and p-coumaric acids. During the fermentation of alcoholic beverages like beer and wine, specific microbial cinnamoyl esterase activities can release free phenolic acids (Couteau et al., 2001). By decarboxylating these free phenolic acids, S. cerevisiae yeast strains and lactic acid bacteria can further metabolize them to create other volatile phenolics. Ferulic and p-coumaric acids can be decarboxylated to produce the equivalent 4-vinylguaiacol and 4-vinylphenol, then reduced to produce 4-ethylguaiacol and 4-ethylphenol, respectively. The concentration and type of food both affect how these volatile phenolics affect flavor. At low concentrations, they may enhance flavor. High concentrations of these volatile phenolics, however, give alcoholic beverages a medical phenolic off-flavor. Candida versatilis, not Z. rouxii, is the yeast that ferments ferulic and p-coumaric acids to produce these crucial taste components. LAB naturally grows in a wide range of settings where soluble carbohydrates, products of protein breakdown, and low oxygen tension are present. Many LABs produce extracellular polysaccharides, which help them attach to surfaces and function as biofilms. LAB generally have some resilience to increased osmotic pressures and are robust to low pH environments. When it comes to the commercial usage of LAB, where bacteria are crucial to the preservation and microbiological safety of many fermented foods, these properties are crucial. Bacteriocins are tiny molecular mass peptides produced by ribosomes and released extracellularly. They can kill or restrain bacteria of the same species or different genera. Even thought nisin is now the only purified bacteriocin utilized in industrially processed foods, several bacteriocins generated by diverse LAB may have future uses in food items.

5.9 LAB

Lactic acid bacteria (LAB) have antifungal capabilities and can be utilized to stop the growth of mycotoxins (Oliveira et al., 2014). By adjusting the intrinsic and extrinsic environments, LAB can successfully compete with undesirable microbes like Staphylococcus aureus. This competition may be observed in the process of making sauerkraut, where one of the phases involves adding salt to reduce water activity while also removing moisture and nutrients from the chopped cabbage. Now, LAB growth is preferred to that of endogenous Gram-negative bacteria. Gram-negative bacteria are further made to cease growing and gradually disappear due to the subsequent drop in pH brought on by the formation of lactic (and acetic) acid by LAB. Studies have shown that LAB increases the solubility (bioavailability) of important metals including iron, manganese, and zinc but not copper during the fermentation of plant material such as carrot juice. LAB degrades oxalate, phytate, and tannins during this process. LABs can be found naturally in a variety of conditions, including the surface of plant materials that are in the process of growing and after harvest. The ability to exploit the available nutrients, tolerance of the relevant environmental conditions (intrinsic and extrinsic factors), and possibly the synthesis of antimicrobial chemicals to ensure survival are just a few examples of how this adaptation can manifest itself. The most obvious thing to look for when selecting an appropriate LAB for biopreservation and fermentation purposes is strains or species that have previously adapted to thrive well under the specified plant nutritive conditions.

Leuconostoc, Lactobacillus, Weissella, Pediococcus, Enterococcus, and Lactococcus are the LAB genera most frequently linked to fruits and vegetable products in both culture-dependent and independent methods. Species like Leuconostocmesenteroides and Leuconostoccitreum predominate in the heterofermentative LAB. LAB isolated from sauerkraut fermentations were identified using a rRNA gene intergenic transcribed spacer-polymerase chain reaction technique together with 16S rRNA sequencing. The molecular approach revealed the presence of strains like Leuconostoccitreum, Leuconostocargentinum, Lactobacillus paraplantarum, Lactobacillus coryniformis, Weissella sp., and the newly described Leuconostocfallax, indicating that the LAB flora developing during the fermentation of sauerkraut is significantly more diverse. Leuconostoccitreum was discovered to predominate in the early phase of fermentation while Lactobacillus sakei/Lactobacillus curvatus and Lactobacillus brevis dominated in the latter phases. Plant material fermentation is a traditional preservation technique that probably has its roots in Asia. Kimchi, a traditional Korean meal, is currently the most widely consumed fermented vegetable dish in many Asian countries. Sauerkraut, cucumbers, and olives are popular fermented vegetable products in Europe and North America. The consumption of fermented vegetable products ensured continuous access to essential vitamins, including vitamin C, minerals, and dietary fibers for people in tropical climates with just one annual growth season before the advent of modern food preservation technologies like refrigeration, controlled storage, and freezing.

5.10 KIMCHI

Kimchi is a traditional Korean fermented vegetable food that is often made with Chinese cabbage as the major component and a variety of other ingredients, including fish sauce, radish, ginger, and garlic (Kim & Chun, 2005). Depending on the type of kimchi, the materials used, and whether it is produced domestically or industrially, many kimchi preparation techniques are utilized. The types of kimchi vary with the region due to variations in weather. The ingredients, salt and spice additions, fermentation, and condiments affect the flavor. Pretreatment, brining, ingredient blending, and fermentation processes make up the main procedure. We need to cut 3 to 5 cm cubes of cabbage. Pretreatment of Chinese cabbage can be done using a salt solution for 2 to 7 hours before being rinsed with fresh water and drained. Sliced radish, carrots, green onion, onion, garlic, ginger, red pepper, salt-pickled seafood, and dried salt are mixed to create a premixture in a separate process following the recipe. The processed cabbage and this premix are mixed well. This combination is then put into a fermentation jar that is covered and let to ferment for one to three weeks at low temperatures of two to ten degrees or for one to three days at twenty to twenty-five degrees. Following fermentation, kimchi is typically kept between 0° to 8°C and served chilled.

The complex microflora of the raw material contains naturally occurring LAB and yeast strains, which are primarily responsible for the fermentation. Leuconostocmesenteroides start the fermentation of kimchi in anaerobic conditions. This organism varies from other LAB species in that it can withstand rather high salt and sugar concentrations. Over a wide range of temperatures and salt concentrations, Leuconostocmesenteroides stimulate vegetable growth more quickly than any other LAB. Due to its heterofermentative metabolism, it quickly lowers pH and produces acids and carbon dioxide that prevent the growth of harmful microbes. Lactobacillus can grow in anaerobic conditions created when carbon dioxide replaces oxygen in the air. In addition to stabilizing the ascorbic acid (vitamin C) present in fermented vegetables, the removal of oxygen aids in maintaining vegetable color. Leuconostocmesenteroides become substantially inhibited as the pH falls to between 4.6 and 4.9, while other LAB, including Streptococcus, Lb. brevis, Pediococcus cerevisiae, and Lb. plantarum, continue to ferment the material. Denaturing gradient gel electrophoresis, a molecular technique based on the 16S rRNA gene and culture independence, was used to analyze the kimchi microflora. They discovered that the main microorganisms responsible for kimchi fermentation were W. confusa, Leuconostoccitreum, Lb. sakei, and Lb. curvatus. The genome-probing microarray for quantitative, high-throughput monitoring, polymerase chain reaction-enzyme-linked immunosorbent assay for the screening of bacterial isolates from fermented vegetables and sodium dodecyl sulfate-polyacrylamide gel electrophoresis profiles of whole cell proteins for the identification of distinct microbial communities are additional techniques that have been used.

The endogenous plant bacteria, salt concentration, fermentable carbohydrates, other nutrients, the presence of inhibitory substances as well as oxygen, pH, and temperature are the most important parameters that influence kimchi fermentation. Temperature, pH, and salt content have an impact on how quickly and how much lactic acid is fermented. Since using well-defined starter cultures can increase the control and predictability of the fermentation process, it is of great commercial interest. Although the kimchi fermentation pattern is comparable to other vegetable lactic acid fermentations, there is a higher degree of variation in the associated LAB flora than in other products because of the variety of kimchi items. The brining process is crucial for the fermentation of kimchi, just like it is for other vegetable fermentations. Brining inhibits the growth of some unwanted bacteria that may ruin the kimchi while osmotically extracting water and nutrients from the raw ingredients. By raising the salt level of the Chinese cabbage or radish, it also creates conditions that are generally favorable for LAB. After bringing Chinese cabbage, the LAB count increases. There is a noticeable reduction in the numbers of bacteria, yeasts, and molds following the brining treatment. The genera Lactobacillus, Leuconostoc, and Pediococcus are responsible for the LAB that predominate in Western vegetable fermentations like sauerkraut. Lactobacilli homofermentative strains yield 85% lactic acid from glucose, whereas heterofermentative strains yield equimolar amounts of lactic acid, carbon dioxide, ethanol, and acetic acid. The phenotypically similar strains of Lb. sakei and Lb. curvatus are thought to be advantageous for sauerkraut production.

Research on the ecology of commercial sauerkraut fermentations using conventional biochemical identification methods identified the main microorganisms as the obligate heterofermenter Leuconostocmesenteroides, the facultative homofermenterLb. plantarum, the homofermenter Pediococcuspentosaceus, and the obligate heterofermenterLb. brevis. Several LAB species are present in the fermentation of sauerkraut, including Leuconostoccitreum, Leuconostocargentinum, Leuconostocfallax, Lb. paraplantarum, Lb. coryniformis, and Weissella sp. Along with the beneficial LAB, cabbage also contains harmful microorganisms that might obstruct the production of sauerkraut. The effectiveness of the fermentation process control over unfavorable microbes has a significant impact on the product quality. Salt is added to shredded cabbage that has been put in a container. The cabbage juice, which includes nutrients good for LAB activity and fermentable carbohydrates, can be removed mechanically. The initial fermentation is initiated by the heterofermentative gas-producing cocci. The activity of this group of LAB slows down when the lactic acid content reaches 0.25 to 0.3 percent, although their enzymes continue to work (Mullan, 2000). The lactobacilli continue fermentation until the acidity level rises to 1.5 percent to 2 percent. These bacteria are inhibited by the high salt content and the cool temperature. Lactic acid, along with minor amounts of acetic and propionic acids, a mixture of gases, a small quantity of alcohol, and a mixture of aromatic esters, are the by-products of the fermentation process for sauerkraut. Esters are produced when the alcohol and acids combine to give sauerkraut its distinctive flavor. The acidity contributes to the longer shelf life by limiting the growth of organisms that cause deterioration.

5.11 SAUERKRAUT

One of the crucial elements in the production of sauerkraut is temperature management. 21°C is the ideal temperature for sauerkraut fermentation. Due to the development and metabolism of Leuconostocmesenteroides, an ideal temperature range for starting fermentation is between 18 and 22 degrees Celsius. In contrast, Lactobacillus species thrive best at temperatures over 22°C. Initiating the sauerkraut production process and having an impact on the final product quality are both crucial roles that salt provides. Salt is applied to the cabbage to eliminate the juice, which improves the conditions for the growth of the desirable LAB flora. Salt is added until the final concentration is between 2.0% and 2.5%. Cocci is unaffected at this dose; however, lactobacilli are inhibited. Pure sodium chloride must be used because salts with additional alkali may neutralize the acid. The use of starter cultures can enhance the fermentation process to assure the safety and health aspects of the products rather than relying on the endogenous LAB flora and intrinsic heterogeneity across different cultivars, processing situations, etc. An important benefit of using starting cultures is that they produce sauerkraut of a higher quality and hasten the fermentation process. Starter cultures also aid in achieving the proper level of acidity in the fermented product, which inhibits unwanted microbes. Starter cultures like Lactococcus lactis can be added without having a negative impact on the final quality. The natural succession of microorganisms is not disrupted by these starters because they only live for a brief period of time yet start the acidification process. Starter cultures for sauerkraut are made from racemase-deficient Lactobacillus bavaricus strains (Hui & Evranuz, 2012). A stable product with the unique flavor and aroma of sauerkraut (Figure 5.3) requires proper sequencing of the microbes. It should be noted that starter cultures for subsequent fermentations of sauerkraut were done using the juice from prior fermentations. The effectiveness of using old juice is heavily influenced by the types of organisms it contains and how acidic it is.

Figure 5.3. Sauerkraut.

(**Source:** *Von Bdubay - Eigenes Werk, CC BY-SA 3.0, https://commons.wikimedia.org/w/index.php?curid=15459964, Creative Commons License)*

5.12 SOIL MICROBES

The aerobic soil microbes that break down the protein to cause unfavorable flavor and texture changes are responsible for sauerkraut deterioration.

The rate of fermentation, as well as the regulation of temperature and salt concentrations, can be used to control the growth of these aerobes. When the regular pattern of bacterial development is disrupted, the end product typically has a soft texture. The softening of the cabbage appears to be caused by Lactobacillus sp., which breaks down cabbage tissue more than the cocci. The development of Lactobacillus sp. is aided by high fermentation temperatures and low salt concentrations.

The Lactobacillus sp. will develop quickly at the first stage of fermentation if the initial salt concentration is too low. The development of microorganisms can be influenced by a number of factors. For instance, uneven salt distribution and a lack of adequate cabbage juice to completely cover the sauerkraut throughout the fermentation process might encourage the growth of unfavorable aerobic bacteria and yeasts, which results in the development of bad odors and discoloration.

The development of dark patches could occur if the fermentation temperature is too high due to the proliferation of unfavorable pigmented bacteria. If the salt is applied in an excessive amount or unevenly distributed, the group of yeasts that cause pink sauerkraut will produce a bright red pigment in the juice and on the surface of the cabbage. This will allow the yeast to thrive while inhibiting the LAB.

The concept of employing harmless commensal bacteria in biopreservation to prevent the growth of foodborne pathogens and spoilage organisms is not new, but it has recently been studied in relation to meat and fish products. This intentional use of microbes and their metabolites attempts to increase food safety and shelf life.

The utilization of active LAB cultures for biopreservation and their metabolites, such as bacteriocins, have drawn a lot of attention due to their antibacterial and antifungal properties. Consumers are increasingly in need of fresh-cut, ready-to-eat fruit and vegetable items; therefore, the food sector needs to find solutions for managing the microbiological risks and quality issues that come with them. An appealing alternative to chemical preservatives may be biopreservation using LAB of plant origin. Using the indigenous knowledge of biopreservation, perishable and seasonal green vegetables, radishes, cucumbers, and young edible sensitive bamboo shoots, are traditionally fermented into edible items. Wheat and barley are consumed across the Indian subcontinent as well as in western and middle eastern Asia. The Indian subcontinent is the only place where ethnic fermented cereal items are typically produced.

5.13 FERMENTED CEREALS

In Asia, fermenting cereals into a variety of regional alcoholic beverages is common (Ashaolu & Reale, 2020). The two prevalent species of tea are 1. Camellia sinensis var. sinensis and 2. C. sinensis var. assamica, and tea is the 2nd most consumed beverage in the world after water. Although many drink black tea, several ethnic Asian communities have their unique fermented teas, like kombucha from China, puer tea, and miang from Thailand. Ethnic alcoholic beverages have a ritualistic significance among the ethnic people of Asia where social events call for the provision and consumption of alcohol in addition to being a recreational beverage. It's interesting that Asia has little or no experience with the malting process used to make alcohol. In several Asian nations, the fermentation of alcoholic beverages traditionally uses dry mixed starters containing a consortium of microbes. Three different kinds of amylolytic mixed cultures or inocula are typically employed in Asia to transform grain starch into sugars, which are then converted into alcohol and organic acids. Type I is a conglomerate of mycelial or filamentous molds, amylolytic and alcohol-producing yeasts, and lactic acid bacteria (LAB) with rice or wheat as the substrate, forming dry, spherical, or flattened balls of varied sizes. With material from an earlier batch, the starter is inoculated. This mixed flora is dried and used to produce alcohol or fermented dishes from starchy ingredients after a minimal period. Type II is popular in Japan, a starter known as koji is used to create alcoholic beverages, notably saké, by combining Aspergillus oryzae with Aspergillus sojae. While proteases are created to digest the soybean protein, koji also creates amylases that break down starch into fermentable sugars that are then used for the second stage of yeast fermentation to create the non-alcoholic fermented soybean products known as miso and shoyu.

Type III includes whole wheat flour. It is wet and formed into substantial, dense cakes that are then incubated in filamentous molds and culture yeasts to make alcohol from the fermentation of starchy materials. China uses this kind mostly to produce alcohol. Among Asians, the Indian and Chinese communities have arange of ethnically-based fermented dishes and alcoholic beverages. Fermented vegetables, fermented bamboo shoots, fermented legumes, fermented cereal, fermented tea, and fermented beverages are the types of plant-based fermented foods found in Asia. Gundruk is a traditional fermented vegetable from the Himalayan regions of India, Nepal, and Bhutan. Fresh leaves of the indigenous vegetable rayo-sag (Brasiccarapa subspecies campestris variation cuneifolia), mustard, and cauliflower are used to make gundruk. This combination is then mashed, dried, and shredded before being placed into an airtight clay jar or container. The container is stored in a warm environment and lets 7 to 10 days to naturally ferment. In contrast to kimchi and sauerkraut, freshly fermented gundruk is sun-dried for three to four days before eating, whereas dried gundruk can be kept for more than two years at room temperature. Gundruk is consumed as a pickle or in soup. Gundruk has been used to isolate the bacteria Lactobacillus fermentum, Lactobacillus plantarum, Lactobacillus casei, Lactobacillus casei subsp. pseudoplantarum, and Pediococcuspentosaceus. Some LAB isolated from gundruk exhibited probiotic characteristics, acidification, antibacterial capabilities, and the capacity to break down antinutritive components. Gundruk, which is regarded as a tasty appetizer, includes organic acids such as lactic, acetic, citric, malic, and acetic acids. The primary flavoring agents in gundruk are cyanides and isothiocyanates, which are followed by alcohols, esters, and phenyl acetaldehyde. Increases in the free amino acids glutamic acid, alanine, leucine, lysine, and threonine as well as palmitic, oleic, linoleic, and linolenic acids have been seen in gundruk.

Kimchi, one of the most well-liked ethnic fermented vegetables in Korea, is a hallmark of the country's rich culinary tradition. In Korea, the term "kimchi" is used to refer to a variety of food that has been fermented with cabbage, radish, and garlic. Kimchi is consumed daily in amounts between 50 and 200 g per person. The components used to make kimchi are Chinese cabbage, radish, cucumber, red pepper flakes, garlic, ginger, and green onions. In the traditional way of making kimchi, chopped cabbage and radish are combined with chopped hot red pepper, garlic, leaf mustard, and some seaweed after being salted. The combination is then fermented in a covered clay pot. The kimchi is removed and served as a side dish. Whole cabbage kimchi (jeotgukji), diced radish kimchi (kakdugi), and water kimchi are the three main varieties of kimchi that are frequently made in Korea (yeolmu). Leuconostockimchii, Leuconostocmesenteroides, Leuconostocpseudomesenteroides, Leuconostoc lactic, Lactobacillus brevis, Lactobacillus kimchii, Lb. plantarum, Lactococcus inhae, Weisella kimchi, Weisellakoreensis, and Tetragenococcuskoreensis were used. Several non-LAB and yeast species, including Halococcus, Natronococcus, Natrialba, Haloterrigena, Lodderomyces, Trichosporon, Candida, Saccharomyces, Pichia, Sporisorium, and Kluyveromyces, have also been identified in kimchi. Kimchi is a functional food that provides a number of advantages for your health. It possesses probiotic qualities, antioxidants, etc. Consuming kimchi was shown to prevent constipation and colon cancer, lowers blood cholesterol, inhibit atherosclerosis, have anti-stress effects, and lower the risk of depression, osteoarthritis, and liver disease. Due to its flavor and usefulness, kimchi is no longer just a staple of the Korean diet and is now frequently consumed by non-Koreans around the globe as a salad.

5.14 SINKI

In the Himalayan regions of India, Nepal, and Bhutan, sinki is a traditional fermented radish dish made through pit fermentation (Tamang et al., 2005). In the Himalayas, lactic acid fermentation is used in a special way to bio-preserve perishable radishes known as pit fermentation. In a dry area, a pit of about 2 to 3 feet in diameter is dug, cleaned, plastered with mud, and heated by burning. The pit is lined with bamboo sheaths and paddy straw after the ashes have been removed. The taproots of the radish are dried for two to three days, then crushed, squeezed, and pressed into the pit. They are then covered with dried leaves and supported by large stones or planks. The pit's top is coated with mud and allowed to organically ferment for 22 to 30 days. Sinki that has just finished fermenting is removed, diced, and sun-dried for three to five days, then kept at room temperature for future use. Dry sinki (Figure 5.4) can be stored in an airtight container for a number of years. It is consumed as a pickle or in soup. Leuconostocfallax, Lb. plantarum, Lb. brevis, Lb. Casei have been isolated from sinki.

Figure 5.4. Sinki.

*(**Source:** https://upload.wikimedia.org/wikipedia/commons/4/42/Sinki_ko_jhol_%28Fermented_radish_soup%29.jpg, Creative Commons License)*

5.15 SUNKI

Sunki is a traditional Japanese fermented vegetable dish prepared from the leaves and stems of the red turnip (Tomita et al., 2018). Turnip is boiled, then allowed to ferment for one to two months after being inoculated with zumi (wild apple). Sunki is consumed with miso soup and rice. Sunki has yielded isolates of P. pentosaceus, Lb. plantarum, Lb. brevis, Lactobacillus buchneri, Lactobacillus kisonensis, Lactobacillus otakiensis, Lactobacillus rapi, and Lactobacillus sunkii. Khalpi is a Nepalese and Indian ethnic fermented cucumber product. Ripe cucumbers are sun-dried for two days before being placed in a covered bamboo jar to spontaneously ferment for three to five days. The pickle form of khalpi is eaten. The four main LABs in khalpi fermentation are Leuc. fallax, P. pentosaceus, Lb. brevis, and Lb. plantarum. Goyang is a traditional fermented wild plant cuisine typically made by Sherpa women in the high mountains of India and Nepal's Himalayan area. The leaves of the wild edible plant (Cardamine macrophylla) are gathered, cleaned, and cut into pieces. The excess water is squeezed out, and the pieces are then securely packed into bamboo baskets that have two to three layers of fig plant leaves inside of them. The baskets are then covered with leaves from the fig shrub and let to naturally ferment for 25 to 30 days at room temperature. Goyang that has just finished fermenting is moved to an airtight container that can be kept for two to three months. Goyang's shelf life can be increased by shaping freshly fermented goyang into balls and letting them sun dry for two to three days before storing for a few months. Goyang is consumed in soup form. Goyang has yielded isolates of L. plantarum, L. brevis, Lactococcus lactis, Enterococcus faecium, P. pentosaceus, and yeasts from Candida spp.

5.16 PAO CAI

Similar to kimchi from Korea, pao cai is a Chinese ethnic fermented cabbage dish. In northeast China, a dish identical to pao cai known as suancai is prepared and consumed (Wu et al., 2015). In China, there are about 11 different primary forms of fermented or pickled vegetables. Pao cai is a morning dish that is eaten with congee and is sweet and sour rather than spicy. Pao cai is produced in a variety of ways and with varying degrees of flavor all around China. Vegetables are chosen, prepped, cleaned, drained, and chopped into pieces during manufacturing. Seasonings (sugar, salt, vinegar, and alcohol) are then added before the vegetables are sealed in jars with water and let to ferment for a week or more. Pao cai is produced in

China more than 2 million tonnes per year, earning 150 million Chinese Yuan yearly. The functional LAB in pao cai are Leuc. mesenteroides, P. pentosaceus, Lactobacillus pentosus, Lactobacillus plantarum, Lactobacillus brevis, Lactobacillus fermentum, etc.

5.17 FU-TSAI

The Hakka tribes in Taiwan produce the traditional fermented mustard foods known as fu-tsai and suan-tsai (Chao, Wu, Watanabe, & Tsai, 2009). Freshly produced mustard is dried, layered in a bucket with 4 percent salt, airtight packed, and let to naturally ferment for 7 to 2 months at room temperature. The fermented product is referred to as suan-tsai if fermentation is allowed to continue for around two months. To make fu-tsai, however, the partially fermented mustard is taken out of the bucket after seven days, washed with water, and then dried in the sun for two days before being fermented once more for at least two days. This process of fermenting at night and sun-drying in the morning is done two or three times. After being cut into bits and packed tightly into glass bottles, pottery pots, or plastic containers, the partially dried mustard is sealed and left to mature for three months upside down. The end outcome is known as fu-tsai. Suan-tsai and fu-tsai are consumed as stewed or fried dishes with shredded beef. P. pentosaceus, Tetragenococcushalophilus, Lactobacillus farciminis, Leuconostocmesenteroides, Leuconostocpseudomesenteroides, Weisellacibaria, and Weisellaparamesenteroides are isolated from fu-tsai. Jeruk is a fermented vegetable native to Malaysia that is made from everyday fruits and vegetables. Pickled fruit is consumed by youngsters and expecting moms because of its sweet-sour flavor. Pickled vegetables are prepared similarly to fresh vegetables. Jeruk contains species of Leuconostoc, Lactobacillus, Pediococcus, and Enterococcus. An ethnic fermented vegetable preparation made from mustard leaves is called Pak-gard-dong in Thailand. Black mustard leaves are gathered, cleaned, and allowed to wilt in the sun before being added to and packed into a container for 12 hours. After the water is removed, 3 percent sugar is added and fermented for 3 to 5 days at room temperature. The three main LABs responsible for the fermentation of pak-gard-dong are P. pentosaceus, Lb. brevis, and Lb. plantarum.

5.18 PAK-SIAN-DONG

In Thailand, Pak-sian-dong is a traditional fermented leafy vegetable product. It is made from Pak-Sian plant leaves (Gynandropis pentaphylla) (Ekpong,

2009). After being completely washed with water, fresh vegetables are laid out in the air or out in the sun to dry out until the sample is noticeably flaccid. After that, it is combined with water, salt, and sugar and preserved for two to three days in a tightly closed container. In Pak-sian-dong, you can find Leuconostocmesenteroides, Lb. fermentum, Lb. buchneri, Lb. plantarum, Lb. brevis, and P. pentosaceus. Sayur asin is a traditional Indonesian food made from fermented mustard cabbage. Dried mustard cabbage leaves are pressed with salt ranging from 2.5 percent to 5 percent. To ensure that enough acid is formed during fermentation, liquid from boiling rice is added. This provides fermentable carbohydrates. Leuconostocmesenteroides and Lactobacillus confusus start the fermentation process. Later, L. plantarum and P. pentosaceus dominate.

CHAPTER

6 Distillation and Fermentation

Contents

6.1 INTRODUCTION

Winemaking (Figure 6.1) starts from picking the grapes to bottling the finished product (Pretorius, 2000). Grapes must be free from bacterial and/or fungal infections-related abnormalities. The activity of acetic and/or lactic bacteria may cause problems if grapes are damaged during harvest and/or transport. The stem, which is often cut off from the grapes and thrown away, is an example of a lignified vegetable structure with a substantial amount of polyphenols. The grape skin contains polyphenols and fragrance/aromatic chemicals. The pulp contains sugars like fructose and glucose, acids like tartaric, malic, and citric acids, and trace amounts of polyphenols and aromatic chemicals.

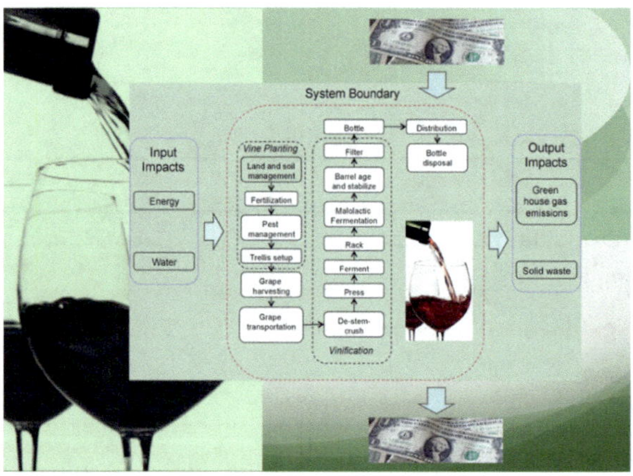

Figure 6.1. Winemaking.

*(**Source:** Zhang C, Rosentrater KA. Estimating Economic and Environmental Impacts of Red-Wine-Making Processes in the USA. Fermentation. 2019; 5(3):77. https://doi.org/10.3390/fermentation5030077, Creative Commons License)*

Grapes are typically processed using a mechanical crusher/destemmer. Fruits are crushed and then moved to the fermentation tank. White wine production requires cautious handling of the grapes since the chemicals in the skin and stem must not enter the must. Yeasts, sulfur dioxide, and nutritional compounds are added after the must has been made. Although yeasts are typically found on grapes and in the winemaking environment, they can have negative effects. Because of this, specific commercial yeast strains are introduced to the must. Due to its inhibitory activity against the natural microflora (bacteria and yeasts) in the juice, sulfur dioxide is employed in winemaking (50–100 mg/L), aiding the action of the selected yeasts added to the must. This also acts to suppress the oxidative enzymes that cause the wine to oxidize. As nutrition for the yeast, ammonium salts and vitamins (biotin and thiamin) are frequently supplied. Sugar may be added if the sugar concentration is low. This addition is, however, strictly bound by local laws. For instance, sucrose cannot be added in Italy. The primary, alcoholic fermentation (AF) begins approximately 10 hours after the yeast inoculation. The yeast cells metabolize the sugars in the must during this fermentation, which typically lasts for about 8 days and produces alcohol and carbon dioxide. The flavor is influenced by the temperature. White wines should be served between 20°C and 25°C, while red wines should be served between 25°C and 28°C. During the fermentation of alcohol, minor products such as glycerol, acetic acid, alcohols, and acetaldehyde are formed. Low levels of acetic acid, increased alcohol concentrations, and acetaldehyde formation are indicators of wine quality.

6.2 WINE

The second fermentation of wine that takes place during storage is known as malolactic fermentation (MLF) (Bartowsky et al., 2015). Due to climate change, the timing of MLF in modern winemaking is moving from spring to autumn (Northern Hemisphere). This phenomenon occurs either after or during alcoholic fermentation. But more than any other microbial change, MLF has an impact on the post-fermentation stages, including aging, stabilization, and potential spoiling. A common genus of bacteria, lactic acid bacteria (LAB) undergo malolactic fermentation and can be found in many foods and drinks. They are present at every stage of the wine-making process. Grapevine leaves, grapes, various pieces of vineyard equipment, and barrels are just a few examples of the surfaces and settings where LAB can be isolated. Low levels of LAB are present on grapes at harvest, although acetic bacteria and yeast are more prevalent. The term lactic acid bacteria first appeared at the beginning of the 20th century to define a heterogeneous group of bacteria that are classified as spherical (cocci) or rod-shaped (bacilli), Gram-positive, catalase-negative, immobile, non-sporulating, anaerobic, aerotolerant producers of lactic acid as a primary metabolite of sugar. Lactic acid bacteria exhibit high morphological and physiological diversity. In terms of the conditional environment of winemaking, Oenococcusoenis is the most researched and well-known species of LAB since it is the primary species carrying out the biological change known as malolactic fermentation.

Pediococcus is typically detected in wines with higher pH levels and after the MLF. The fact that LAB is responsible for a crucial fermentation step in wine does not make it any different from other foodstuffs, but it can also result in modifications that negatively impact the organoleptic aspects of the product. New species of lactic acid bacteria have been found in both musts and wines in molecular identification studies. These include Lactobacillus bobalius, Lactobacillus uvarum, Lactobacillus nagelii, Lactobacillus vini, etc. Between 103 and 104 colony-forming units/mL of LAB are present in the must phase and early stages of AF during the first stages of winemaking. After that, due to adverse conditions like low pH, lack of nutrients, and the presence of ethanol, only a few species can survive in grape must and wine.

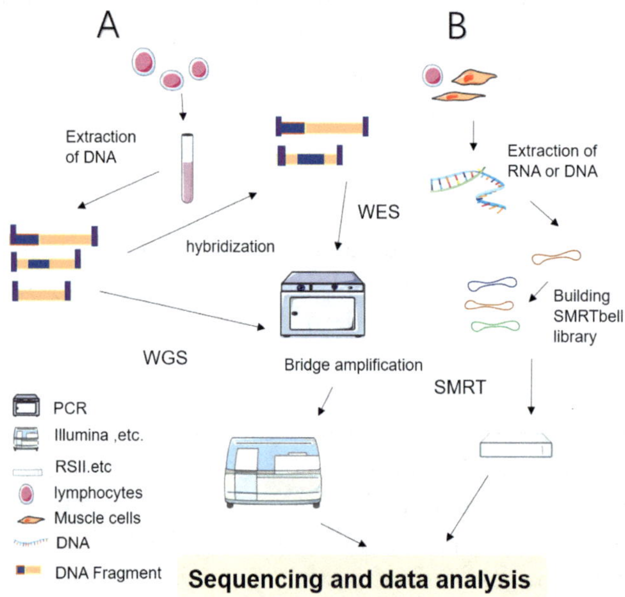

Figure 6.2. Next-generation sequencing.

*(**Source:** Sun H, Shen X-R, Fang Z-B, Jiang Z-Z, Wei X-J, Wang Z-Y, Yu X-F. Next-Generation Sequencing Technologies and Neurogenetic Diseases. Life. 2021; 11(4):361. https://doi.org/10.3390/life11040361, Creative Commons License)*

6.3 NEXT-GENERATION SEQUENCING

Next-generation sequencing (NGS) (Figure 6.2) was used to obtain a thorough understanding of how microbial communities change during grape fermentation and how various fermentation techniques may affect the composition of the microbial community, including bacteria that have negative effects on wine quality (Shin et al., 2018). Depending on the species and strains involved in MLF, bacterial metabolism can vary greatly. In general, all LABs use fermentative metabolism to get energy from carbohydrates since they lack heme-linked cytochromes and catalase. The hexoses i.e., glucose, fructose, mannose, and galactose, as well as pentoses, polyols, and oligosaccharides, can be metabolized by the LAB group. The species and strains involved, as well as the pH, affect this capacity. Since wine LAB cannot utilize malic acid as its only source of carbohydrates, the presence and utilization of fermentable carbohydrates in wine by LAB are crucial for the start of bacterial growth and the occurrence of MLF. Current research has conclusively shown that grape-derived phenolic glycosides considerably promote the development of O. oeni. Malolactic fermentation refers to a decarboxylation reaction rather than a fermentation process because it explains the enzymatic conversion of l-malic acid to l-lactic acid and CO_2 by LAB. Malic acid, an important acid in wine, is changed into l-lactic acid, a monocarboxylic acid, which raises the pH and alters the taste. Wine-derived lactic acid bacteria convert l-malic acid into l-lactic acid and carbon dioxide through a direct reaction, i.e., the formation of the intermediary pyruvic acid. Malate decarboxylase, also known as the malolactic enzyme, is the catalyst for this reaction and needs the cofactors NAD^+ and Mn^{++} to function. All types of lactic acid bacteria isolated from the wine have the malolactic enzyme, which was initially purified using Lactobacillus plantarum.

6.4 SUGARS

Malic and citric acids are not the only energy sources used by LAB. As a result, sugars are needed by malolactic bacteria as a carbon source. Energy (ATP) produced from MLF, however, is helpful for cell growth in situations with limited sugar availability or low pH, which impair sugar metabolism. In wine, LAB initial growth rates and metabolic activity may be stimulated by malolactic enzymes. The malolactic reaction can be carried out by all wine lactic acid bacteria. Oenococcusoeni, however, has been used for a very long time and almost exclusively for the induction of malolactic fermentation in wines, making it the main player that carries out this fermentation. By analyzing the 16S rDNA sequence among LAB, Oenococcus was split off from the genus Leuconostoc. This genus once only contained the species

O. oeni. Oenococcuskitahareae was described in 2006, and more recently, a third species of the genus Oenococcusalcoholitolerans was assigned, isolated from ethanol production processes. Lactobacillus plantarum has been the subject of several scientific investigations in recent years that have also sparked commercial interest in this genus. One of the primary metabolic capacities of LAB in wine is the decarboxylation of malic acid to lactic acid, which is a biological deacidification reaction and of significant commercial value to the winemaking process (Bartowsky, 2005).

The development of methods to encourage microbial deacidification of low pH wines is particularly important since the sensory impact of deacidification is particularly detectable and appreciated in wines with high acidity. Malic acid is naturally present in wines made from grapes grown in cold climates at levels up to 8 g/L, and these wines are better due to lowered acidity. Due to severe conditions (low pH), the induction of MLF in such wines is sometimes challenging to achieve. To advance knowledge in this area and propose and improve techniques like the use of bacterial starters with early inoculation as effective means of achieving MLF, many recent investigations have focused on wines with high acidity. The wines made from grapes grown in warm to hot climates, on the other hand, have lower overall acidity and a further drop in acidity from MLF could negatively affect their quality and result in a plain taste. However, biological deacidification is frequently desirable in wines with high pH levels since it enhances microbiological stability. Even though the absence of acidity brought on by this change is not clearly noticeable or even harmful in certain situations, MLF still has an indirect effect. It is generally accepted that this is done to eliminate nutrients like malic acid, a potential carbon source, and other substances consumed by LAB that might otherwise allow unsuccessful malolactic fermentations or other microbial changes, such as the growth of Brettanomyces in aged wines.

6.5 LAB

The growth and development of LAB in wine, which can have a considerable impact on wine quality, is investigated to be related to a varied range of other metabolic activities in addition to the deacidification response that distinguishes MLF. Wine has a variety of complex aroma and flavor chemicals that are mostly derived from the grape, from yeast metabolism during alcoholic fermentation, and, occasionally, from oak. By producing new compounds and changing compounds acquired from grapes, yeast, and oak during malolactic fermentation, bacteria may contribute to wine flavor. By changing fruit-derived aromas and creating aroma-active chemicals, MLF does affect the final smell and taste. Over the past 15 years, molecular studies have improved our understanding of O. oeni secondary metabolic activities during MLF, which might alter the sensory characteristics of wine. The metabolism of organic acids, carbohydrates, polysaccharides, and amino acids is one of these secondary activities. Glycosidases, esterases, and proteases produce volatile compounds that are above the threshold at which they can be detected by the human nose. Recent research on the genomic diversity of O. oeni strains utilizing array-based comparative genomic hybridization and genome sequencing has confirmed that phenotypic variation between O. oeni strains is essential for creating various wine types. In recent studies comparing the aromatic compounds released from natural precursors by particular Oenococcusoeni strains during malolactic fermentation, some bacteria showed good results and others were the best at preserving the fruity aroma of wines.

6.6 SECONDARY METABOLITES

Bacterial secondary metabolites can have a negative impact on the sensory attributes of wine because many of them are volatile (Tyc, Song, Dickschat, Vos, & Garbeva, 2017). If MLF is uncontrolled, lactic acid bacteria can damage wine during winemaking. They can even spoil wine subsequently, during maturation and aging. The malolactic fermentation metabolic aberrations by the fermenting bacteria (O. oeni or L. plantarum) might impact wine quality. After malolactic fermentation, the proliferation of Lactobacillus, Pediococcus, and even some Oenococcus species in wine can result in a variety of spoiling scenarios because they can produce unfavorable smells and flavor components. The substances responsible for LAB-caused wine spoilage generally range from acetaldehyde through acetic acid, diacetyl, tetrahydropyridines, acrolein, beta-glucans, and biogenic amines. Regarding organic acids, O. oeni and other genera of the lactic acid bacteria break down not only the malic acid but also the citric acid in the wine to produce acetic acid and diacetyl. Diacetyl is regarded as one of the most important tastes created during MLF and is one of the intermediate molecules in the metabolism of citric acid. Diacetyl gives the wine a buttery aroma when it is present in concentrations that are above the threshold. The threshold values in various wines have been shown to range from 0.2 mg/L in Chardonnay wine to 2.8 mg/L in Cabernet Sauvignon. Below 4 mg/L, the chemical can enhance the richness and fragrances of wine, but above this concentration, it develops an unpleasant buttery flavor.

The final level of diacetyl in wine is influenced by several variables, some of which the winemaker may regulate, especially during malolactic fermentation. Diacetyl production is affected by the bacterial strain used, oxygen exposure, fermentation temperature and time, wine type, and sulfur dioxide. It should be noted that oxidative decarboxylation of acetolactate, an unstable intermediate molecule formed during citrate metabolism, results in the formation of diacetyl. Since it is produced as an intermediate metabolite in the reductive decarboxylation of pyruvic acid, its synthesis and degradation by bacteria are directly related to the growth and metabolism of sucrose, malic acid, and citric acid. The utilization of diacetyl by Oenococci has also been reported in earlier studies, which is not surprising given that many LAB contains diacetyl reductase, which turns diacetyl into acetoin and 2,3-butanediol. Tartaric acid can be broken down by lactic acid bacteria strains. This acid catabolism modifies wine, resulting in a minor decrease in acidity and an increase in volatile acidity. One of the sensory carbonyl compounds developed during vinification, acetaldehyde is very volatile and has an apple-like aroma. It accounts for more than 90% of the total aldehyde content in wine and is primarily produced by yeast metabolism.

6.7 ACETALDEHYDE

Acetaldehyde emits a pleasant fruity fragrance when present in small amounts, but when present in large amounts, it produces an unpleasant aroma that has been designated as green, grassy, or apple-like. The addition of SO_2 can hide the aroma, however binding of SO_2 to acetaldehyde lowers the antibacterial and antioxidant properties of the molecule. Acetaldehyde and phenolics combine to produce persistent polymeric pigments that are resistant to SO_2, which enhances the color of red wine (Liu & Pilone, 2000). However, this interaction can also cause phenolic vapor and the eventual deposition of condensed pigments. When SO_2-sensitive strains of O. oeni break down SO_2-bound acetaldehyde, free SO_2 that is produced may cause inhibition and result in minimal MLF. Although the reduction in the acetaldehyde in wine is thought to affect the finished wine, the chemical and sensory effects of the ethanol and acetic acid produced by the degradation of acetaldehyde by lactic acid bacteria are thought to be limited. The amount of SO_2 needed to eliminate acetaldehyde aroma can be reduced by conducting MLF using effective acetaldehyde-degrading strains. Tetrahydropyridines, which are strong nitrogen-heterocyclic chemicals, are formed as a result of the metabolism of ornithine and lysine in wines. Since an increase in pH makes the chemicals volatile, interactions with the mouth/oral tissue lead these molecules to be felt on the rear palate. These substances appear to only be produced by the heterofermentative LAB.

The decarboxylation of the amino acids by microorganisms leads to the production of biogenic amines (BA). Amino acids can be decarboxylated by bacterial species. Given that it causes a rise in pH, this reaction is thought to benefit microbial growth and survival in acidic media. Histamine, tyramine, putrescine, cadaverine, and phenylethylamine are typically present in wine due to the decarboxylation of amino acids, with the first three being the most common. Some authors believe that the presence of these chemicals is a crucial factor that harms alcoholic beverages in terms of both sensory issues and consumer health concerns. Biogenic amines may appear because the malolactic fermentation process in winemaking involves the development of lactic acid bacteria.

Many studies have revealed a wide range of variables that affect the presence of LAB and MLF in wines. Organic acid content, carbohydrates, amino acids, vitamins, and minerals, as well as alcohol content, pH, SO_2, the vinification process, and interactions between LAB and other wine microorganisms, were the most important factors to influence LAB growth, in addition to oxygen and CO_2. While some of the factors affecting MLF feasibility (grape variety, alcohol or potential alcohol, pH, malic acid content) are difficult to change, many others (SO_2, yeast strain selected for AF, temperature, nutrients, etc.) can be controlled by the winemaker to reduce the risk of slow fermentations.

To ensure the effective completion of MLF and to lower the risks connected with spontaneous MLF, commercial starting cultures of LAB were first added to grape must or wine. A delay in the start or completion of MLF, the production of biogenic amines, the development of bacteriophages, and the presence of unidentified/spoilage bacteria are just a few potential risks. All of these factors affect the wine quality. Although MLF can happen spontaneously, winemakers prefer to inoculate with a dependable, readily available starting culture to reduce the risk of a slow MLF. The winemaker can stimulate a beneficial flavor contribution to the wine as well as the quick start and completion of MLF by inoculating a commercial starting culture, most of which contain O. oeni as the sole LAB culture.

6.8 STARTER CULTURES

Even now, the majority of commercial starter cultures still employ O. oeni as their primary species. The majority of cultures are made using a single strain or a combination of two or three strains. Today, there are numerous commercial starting cultures available to promote malolactic fermentation; the majority of these starter cultures are made up of lactic acid bacteria strains with high levels of malolactic activity, high tolerance for low pH, and high levels of ethanol. There are several ways in which these starter cultures have been commercialized, including fresh, frozen, and lyophilized cultures. There has been a lot of research and development since the introduction of malolactic starter cultures for enhancing the induction of MLF to optimize inoculation strategies and strain selection to further increase MLF effectiveness. Starter cultures can be sequentially inoculated or co-inoculated with yeast. Co-inoculation typically speeds up vinification compared to sequential inoculation. This has significant complications in the wine sector: Vinification rates that are accelerated stabilize wine more quickly and lower the chance of spoiling. Since grapes lack some of the limiting components found in wine, it is a more favorable environment for microbial growth than wine. Under these conditions, malolactic bacteria are predicted to adapt and function more effectively. During alcoholic fermentation, the barriers to microbial activity, in particular ethanol and some compounds made by yeast, gradually build up, giving bacteria time to adapt and giving lactic acid bacteria a better chance of surviving. O. oeni is heterofermentative, which could be a possible issue. This indicates that under specific conditions, acetic acid is one of the by-products of its metabolism of sucrose. So, O. oeni may be able to create wines with increased volatile acidity. Given the possibility of simultaneous consumption of sugar and malic acid, the bacteria used in this type of fermentation must be specially adapted to prevent spoiling.

The industrial application of co-inoculation in the production of many red and some white wines has attracted considerable interest on a global scale. Co-inoculation techniques have been shown to aid in the production of a variety of grape varieties that reliably produce malolactic fermentation. Pediococcusdamnosus and other unidentified Lactobacillus bacteria may predominate in many spontaneous MLFs in wines with a pH exceeding 3.5, increasing the risk of organoleptic anomalies. Recently, bacterial strains capable of causing a high-quality MLF in high-pH wines have been isolated through the application of adequate microbiological selection criteria to wines. Using starter cultures other than O. oeni with a focus on Lactobacillus plantarum is one of the new trends in the application of malolactic fermentation. L. plantarum is also ideal for MLF in a co-fermentation or even pre-alcoholic fermentation since it prefers malate as an energy source at low pH and can start decarboxylating malate before a yeast starter culture has been added. Compared to O. oeni, L. plantarum generates a wider variety of extracellular enzymes, including glycosidases and esterases. L. plantarum may improve wine properties more than O. oeni because these enzymes may be crucial in the development of wine sensory properties through the release of flavor molecules. Due to the high-throughput sequencing technologies that are making advancements in this field of research, wine production and grape and wine microbiota exhibit regionally defined patterns associated with vineyard and climatic conditions and are becoming more and more of a focus of research (Miras-Avalos & Araujo, 2021).

It might be difficult to determine the best approach to prevent wine from spoiling. Even cleanliness and wine chemical compositions are ineffective at keeping out undesirable microbes. It can be difficult to stop bacterial growth in wines with lower ethanol concentrations and low acidity, as grapes must is not treated with pasteurization before yeast inoculation, unlike the treatment of wort in beer brewing. Wines with high ethanol concentrations and high wine acidity can inhibit the development of bacterial populations. For instance, improper aeration during and/or after the wine-making process can cause acetic acid bacteria to proliferate and become active. These microorganisms fall under the genera Acetobacter, Acidomonas, Gluconobacter, and Gluconacetobacter. Of these, Gluconobacteroxydans, Acetobacter aceti, Acetobacter pasteurianus, Gluconacetobacterliquefaciens, and Gluconacetobacterhansenii are typically linked to grapes and wine but are easily regulated by limiting oxygen. The main ingredient for the preservation of wines is sulfur dioxide (SO_2). Wine can contain carbonyl and keto chemicals that can bind to SO_2 and reduce its effectiveness, increasing the total amount of SO_2 needed. Production issues could arise as a result of rising customer demand for low-sulfite and organic wines if SO_2 binders are not handled correctly during vinification. To achieve the ideal ratio of free and bond SO_2, the appropriate control of the fermentation stages (alcoholic and malolactic) can be crucial in this situation. Malolactic fermentation in particular has been shown to lower binding SO_2 levels. Microbiological wine stabilization one week after malic acid depletion is an efficient strategy for the maximum removal of SO_2 binders while lowering the risk of potential post-ML spoilage by O. oeni leading to the production of acetic acid and biogenic amines.

6.9 BIOGENIC AMINES

Biogenic amine production is another area where MLF management may be crucial to prevent spoiling (Hernandez-Orte et al., 2008). Previous studies demonstrated that the majority of commercial malolactic bacteria did not produce BA, and the use of commercial malolactic starters in wines helps reduce the BA levels because the concentrations of BA in the inoculated wines were significantly lower than those of the control wines. These findings imply that choosing the right malolactic starters can reduce BA formation. The winemaker is especially advised to inoculate particular malolactic starters to replace the native microflora when BA-producing bacteria are present in it. However, this does not represent the only option when starter cultures' dominance of the local BA-producer microflora is insufficient. MLF control has an impact on how Brettanomyces grow throughout subsequent wine aging. MLF suppressed Brettanomyces growth in wines, resulting in a product with little to no volatile phenols. Wines without MLF or with late spontaneous MLF that progressed slowly allowed Brettanomyces to proliferate, resulting in a product with higher volatile phenols. Early malolactic bacteria inoculation of wine may be a technique for reducing the likelihood of volatile phenol formation. Delaying MLF raises the risk of volatile phenol generation because Brettanomyces can expand during the interval between alcoholic and malolactic fermentation by taking advantage of the lack of SO_2. Thus, it is suggested that using malolactic starters is an excellent strategy to prevent Brettanomyces from growing. Scientific research has demonstrated that co-inoculation or early sequential inoculation is the best method for preventing Brettanomyces contamination by shortening the lag time between AF and MLF. It is advised to remove all microbes following malolactic fermentation, particularly by adding SO_2. The use of chitosan is one option for reducing the growth of undesired bacteria in the event of infection, notably Brettanomyces.

6.10 MICROBIAL ACTIVITY

Microbial activity is involved in the brewing of beer at every stage. Given that beer is the product of conventional food fermentation, the majority of these actions are good. Others, on the other hand, pose hazards to the finished product and need to be carefully managed. With rare exceptions, beer fermentation is a monocultural microbial event, although the entire process incorporates a series of microbial components that have an impact on the finished product. In barley, plant-microbial interactions and the microbiological condition of the grain may have an impact on beer quality. To break dormancy after harvest, barley may be kept for a while before being malted. Microbes are still developing and interacting with the living grain throughout this time. To prevent microbial growth, conditions must be carefully controlled to guarantee that the grain is stored in a low-moisture, low-temperature environment. Fusarium spp. can produce mycotoxins, which survive the brewing process. Mycotoxins such as deoxynivalenol (DON), nivalenol, T-2 toxin, HT-2 toxin, and diacetoxyscirpenol have been found in barley. Gushing is another problem caused by a fungus that has more direct negative effects on the customer (Lowe & Arendt, 2004). This behavior is brought on by hydrophobins, which are hydrophobic peptides found in fungi. When the beer container is opened, these chemicals cause the spontaneous release of gas as they act as nucleation sites for CO_2 bubbles. It is commonly known that gushing and fungus growth are related. Application of LAB in the field is an effective strategy for reducing Fusarium growth on barley. Premature yeast flocculation (PYF) during fermentation can also be due to plant pathogenic factors. When lectin-like glycoproteins on other cells attach to cell wall mannoproteins in yeast, flocculation results. Before the complete reduction of sugar in PYF, yeast aggregation and settling take place, which drastically lowers the quality of the beer.

6.11 MALTING

Steeping, germination, and kilning are the three main processes in the malting process. The repeated steeping and aeration cycles encourage plant growth, and kilning reduces the number of microorganisms, microbial activity during germination may have an impact on the quality of the beer later on. After kilning, low-moisture conditions must be kept to prevent microbiological deterioration, especially because the malt products are still hygroscopic and full of soluble nutrients. To lessen the inhibitory effects of microbial growth on malt quality, the steep liquor should be changed between air chambers, dissolved nutrients should be reduced, suspended biomass should be added again, and the steep temperature should be controlled. Steep liquor should be inoculated to prevent the development of harmful bacteria during germination. When applied during steeping, Wickerhamomyces anomalus has been proven to suppress Fusarium development on malt, preventing the production of hydrophobin and beer gushing. It has been demonstrated that Geotrichumcandidum and Lactobacillus plantarum reduces Fusarium development on malt.

The microbial load decreases during mashing, although thermotolerant microorganisms, particularly homofermentative LAB, continue to function in the nutrient and moisture-rich environment. However, bacterial proliferation might have advantageous effects, and lactic acid bacteria mash acidification can enhance malt extraction, fermentability, and nitrogen output. Froth stability, color, and flavor might be enhanced. Extended mashing can potentially lead to major issues due to bacterial growth. Bacillus spp., for instance, can produce excessive acidification and nitrosamine (by reduction of nitrate to nitrite). High quantities of butyric acid can be produced during Clostridium growth, giving the beer a cheese-like aroma. After the mashing, the wort is sterilized by being boiled for a long time. However, because wort has a high pH (5.5) and is nutrient-rich, it is susceptible to opportunistic spoiling agents once it exits the fermentation tank if the proper precautions are not taken to promote quick fermentation. The most common wort spoilers are gram-negative enterobacteria, particularly species of Klebsiella, Citrobacter, Enterobacter, Obesumbacterium, and Escherichia. These result in the production of 2,3-butanediol, organic acids, and dimethyl sulfide in the wort, which gives beer an unpleasant-aroma. After cooling, oxygenated wort is pumped into fermenters, where strains of Saccharomyces are added to hasten the fermentation of maltose and other sugars into ethanol and carbon dioxide. This results in the quick conversion of the wort into beer. The beer matures, or saturates, with the carbon dioxide produced by the remaining yeast activity during this stage of manufacturing. The stage of brewing beer that takes the longest is maturation. The time it takes for the beer to mature may depend on the recipe that is used, the temperature, and the pressure that is set in the maturation tank. Bottom-fermented beer matures in three weeks or less, light lager in at least thirty days and beer with a high alcohol level (12 percent) in 60 days to several months. A key consideration in developing breweries is the spatial orientation of the tanks (horizontal or vertical). Even if using a horizontal type can save space, vertical solutions are the norm in contemporary breweries, and this design is favored when a lack of room is not a basic constraint. The development of a brewery appears to depend heavily on various cooling arrangements as well as the spatial orientation of the tanks.

Cold air is used to cool uninsulated tanks. Low cost is a plus, but there are some drawbacks, such as high cooling energy requirements. It should be noted that a room with excellent wall and ceiling insulation and a refrigerator door appears to be required for the brewery cellar with single-shell tanks to function as a cooling box. In comparison to the larger cellar with insulated tanks chilled by liquid, the running expenses of cooling are especially high during the warmer months. The additional expenses for a dedicated structure and its maintenance typically outweigh the savings from purchasing a brewery with less expensive technology. Beer Maturation Lagering tanks are more sophisticated and are cooled by glycol or ice water circulated in the double jacket duplicators. These tanks have only positives, such as minimal operating costs for cooling, despite the drawback of the larger original cost. Furthermore, this type of option has few installation requirements, allowing the user to operate comfortably. Each tank may have its maturation control system. Although all of the Saccharomyces strains used to make beer produce ethanol as a by-product of fermentation, they are divided into the ale and lager yeast categories. The more diverse yeasts are the ale yeasts (Saccharomyces cerevisiae strains), which are essentially isolated everywhere. Modern fermenters, however, have hydrostatic pressure that tends to counteract this ale yeast growth, causing the yeast to congregate in the tank's cone. Lager yeast is regarded as "bottom-fermenting" because of its propensity not to rise to the surface under any combination of fermentation conditions (Saccharomyces pastorianus) (Yamagishi & Ogata, 1999). It should be noted that lager yeast is a more complicated organism than ale yeast, and it has been suggested that it evolved through the hybridization of S. cerevisiae and S. bayanus.

6.12 BEER

Beer is an example of an ethanol-water solution with a pH of 4.2 that contains hundreds of distinct molecules dissolved in it. The beer is filtered to get rid of the significant amount of yeast after the fermentation and aging processes are finished. A freshly bottled beer, on the other hand, resembles a closed system because its components are not in chemical equilibrium. Thus, this system will work to achieve the lowest energy and maximum entropy state thermodynamically. Therefore, under storage conditions, molecules are prone to a variety of reactions, which are primarily governed by reaction rates. Several significant aging events are initiated, in part, by oxygen, which accelerates the loss of beer flavor. In its ground state ($3O_2$), oxygen is remarkably stable and does not interact with organic molecules very frequently. Ferrous iron (Fe^{2+}) makes it possible for oxygen to absorb an electron and create the superoxide anion O_2 and Fe^{3+}. The same is true for copper ions (Cu^+ is oxidized to Cu^{2+}). Alcohols, sugars, isohumulones, and polyphenols may function as electron donors in a system of mixed oxidation, which is represented in part by the ions Cu^{+}/Cu^{2+} and Fe^{2+}/Fe^{3+}. The perhydroxyl radical (OOH•), which has higher reactivity than the superoxide anion, can be formed by the protonation of the superoxide anion. Beer pH and pKa indicate that the majority of the superoxide forms in this reaction are perhydroxyl. The peroxide anion, which is easily protonated to hydrogen peroxide (H_2O_2) in beer, can be produced by reducing the superoxide anion by iron/copper ions. Then, by a metal-induced reaction like the Fenton or Haber-Weiss reaction, hydroxyl radicals (OH•) can be generated from H_2O_2 or the superoxide anion (O^{2-}). When beer ages, the concentration of free radicals rises with iron/copper ion concentrations, oxygen concentrations, or storage temperatures. Reactive oxygen species (ROS) interact with a variety of chemical compounds, including alcohols and polyphenols, changing the sensory profile in many ways. One of the most reactive species discovered among these is the hydroxyl radical, which can interact with the ethanol in beer. The 1-hydroxyethyl radical, which is quantitatively the most important radical in beer, was formed when ethanol and the hydroxyl radical reacted.

6.13 ALCOHOLS

The important alcohols in beer are ethanol, 2-methyl-propanol, 2-methyl-butanol, 3-methyl-butanol, and 2-phenyl-ethanol. Due to exposure to oxygen, the associated aldehydes increase. Aldehydes may, however, be present in higher quantities when other factors like high temperatures, low pH, and alcohol are present. Aldehydes may be produced by the amino acids in preserved beer. In the presence of oxygen, leucine or valine boosted the synthesis of 2-methyl propanal and 3-methyl-butanal in beer. The reaction between amino acids and dicarbonyl compounds has been proposed as the mechanism for this reaction. Transamination precedes the subsequent decarboxylation of the resultant ketoacid, which produces an aldehyde. By the Maillard process, more dicarbonyl compounds could be formed in beer. Aldol condensation of carbonyl compounds may be achieved in mild storage conditions. Bitterness can be reduced via the breakdown of acids. Certain acid by-products, like iso-acids, are responsible for the development of taste. The trans-isomer appears to be more sensitive than the cis-isomer during the rapid breakdown of iso-acids in the presence of ROS. Iso-acids can release electrons in the presence of adequate electron acceptors without necessarily involving oxygen species. Different carbonyl compounds with varying chain lengths are produced as iso-acid degradation products, including 2-alkanones (C_3 to C_{11}), alkanals (C_2 to C_{10}), 2-alkenals (C_4 to

C_7), and 2,4-alkedienals (C_6 to C_7). Beer made from malted barley includes just two lipid substrates: linoleic acid and linolenic acid. Barley and malt lipases are responsible for the release of these acids from triacylglycerols. During mashing, these molecules primarily undergo hydrolysis to yield fatty acids. Auto-oxidation or enzymatic oxidation using lipoxygenases are the two oxidation pathways (Ito et al., 2016).

Maillard reactions are defined as intricate chemical reactions involving reducing sugars and proteins, peptides, amino acids, or amines. This kind of reaction involves some heterocyclic substances that can be found in old beers, including furfural and 5-hydroxymethylfurfural. The latter is one of the important heterocyclic compounds in terms of quantity. A carbonyl group from a hexose sugar and an amino group combine to generate a Schiff's base, which then transforms into an imine. This is converted into a more stable 1-amino-1-deoxy ketose through a rearrangement. This product, however, is vulnerable to enolization because of the pH of beer, which results in the release of an amine and the formation of 3-deoxy-2-hexosulose (3-DH). Among other secondary products, this reactive dicarbonyl molecule might result in 5-hydroxymethyl-furfural (HMF). Similar to this process, pentose is used as the starting point for the production of furfural. HMF, furfural, and other substances are Maillard reaction intermediates. They go through additional processes like condensation, dehydration, cyclization, and isomerization, which results in the formation of high-molecular-weight brown pigments (melanoidins). Substantial amounts of chemical condensation between ethanol and organic acids take place while beer is in storage. Ethyl 3-methyl-butyrate and ethyl 2-methyl-butyrate are frequently produced from 3-methyl-butyric acid and 2-methyl-butyric acid, respectively. On the other hand, some esters, like iso-amyl acetate, can be hydrolyzed, which lessens their impact on the flavor of beer. Phenolic polymers interact with proteins during beer storage to create insoluble complexes. Free radicals and ROS can easily interact with polyphenols. Simple polyphenols either through acid catalysis or through oxidative processes polymerize into tannins. For polyphenols to interact with other chemicals, they must first undergo oxidation to quinones or semiquinones. The formation of ethyl bonds between flavanols by acetaldehyde might trigger polymerization processes.

The majority of the antioxidants like ascorbic acid, sulfhydryl compounds, chelating agents, polyphenols, and Maillard reaction products might also have pro-oxidant properties. Melanoidins have been shown to bind metals, serve as reducing agents, and scavenge free and active oxygen. However, reports of its pro-oxidant action have also been made. When compared to the concentrations of an oxidizable substrate, an antioxidant is a substance that considerably slows down or stops the oxidation of that substrate. When compared to sulfite, the phenolic acid scavenging action at beer pH is quite less. However, at low concentrations, some antioxidants, including (+)-catechin, could act as pro-oxidants. Sulfite is a strong flavor degradation inhibitor. It can stabilize the flavor by acting as an antioxidant and a carbonyl scavenger. Radicals can efficiently develop in beer during storage. Strong antioxidant organisms are yeasts. Saccharomyces cerevisiae can reduce a wide range of carbonyls and appears to be crucial in the process of reducing inhibiting chemicals, including their precursors. The oldest known cooking ingredient in food preservation is vinegar. When ethanol is fermented, an acidic liquid called vinegar is created. Acetic acid is a fundamental component of this process. For table vinegar, the concentration normally ranges from 4 to 8 percent by volume, while for pickling, it can reach 18 percent. In natural vinegar types, tartaric acid, citric acid, and other acids are frequently found in trace levels. Its main applications are salad dressings, marinades, and the preservation and flavoring of food. The secondary use for vinegar is as a cleaning agent. There are numerous varieties of vinegar, which are often categorized according to the primary ingredient. There are numerous types of vinegar available on the market, including malt, wine, apple cider, balsamic, and fruit.

Red or white wine is used to make wine vinegar (Budak & Guzel-Seydim, 2010). In the countries of the Mediterranean and Central Europe, this is the vinegar that is most frequently used. There is a wide difference in quality, much like wine. High-quality vinegar is aged in wood for up to two years, giving them a deep flavor. It has less acidity than white or cider vinegar. Typically, vinegar is made from beer in the United Kingdom, Germany, Austria, and the Netherlands. Although its taste varies depending on the specific type of beer from which it is prepared, it often has a malty flavor. Similar to wine vinegar, just one type of fermentation occurs during the manufacture of beer vinegar: the transformation of ethanol into acetic acid. Other vinegar is categorized under a production method that includes two fermentations: one for ethanol synthesis and another for acetic acid conversion. These are produced from fruit wines, typically without seasoning. Common tastes of this vinegar type include apple, raspberry, quince, and tomato. Usually, the end product retains the flavors of the original fruits. The majority of fruit vinegar is made in Europe, where there is an increasing market for premium vinegar made only from certain fruits. However, Asia also produces some variants. In China, jujube vinegar, wolfberry vinegar, and persimmon vinegar are all popular products. Indian rose apple vinegar, known as jamun sirka, is thought to provide medical benefits for the stomach. Balsamic vinegar is a fragrant, aged variety of vinegar that is historically produced in the Italian regions of Modena and Reggio Emilia. White grape must, or concentrated juice, is used in its production. It has a dark brown appearance and a rich, sweet flavor. A series of barrels constructed of different species of wood, such as oak, mulberry, chestnut, cherry, juniper, ash, and acacia, are used to age the wine. The lengthy and intricate aging procedure results in an expensive final product. But in recent decades, a less expensive variety of balsamic vinegar gained popularity and availability. To make the commercial balsamic vinegar that is sold in stores, concentrated grape juice is often combined with strong vinegar, sugar, and caramel.

Malt vinegar can be produced by malting barley, which transforms the grain starch into maltose. The maltose is used to brew successive batches of ale, which is then allowed to ferment into vinegar and aged. It has a normal light brown appearance. East and Southeast Asia are where rice vinegar is most widely used. There are light yellow, red, and black versions of it. The Japanese choose a light style when making salad dressings and sushi rice. Red yeast rice has traditionally been used to color red rice vinegar. In China, black rice vinegar, which is made from black glutinous rice, is very popular. White rice vinegar has less acidity.

It is also produced in France and the United States, cane vinegar, which is made from sugar cane juice, is most popular in the Philippines. Its color varies from deep yellow to golden brown, and its flavor is mild and somewhat similar to rice vinegar. In general, producing vinegar enables the use of overripe fruits and cane by-products like molasses and bagasse. Sugarcane juice can be used to make organically fermented vinegar.

There are a variety of vinegar that may be found all over the world, including coconut vinegar, palm vinegar, raisin vinegar, honey vinegar, and sugar cane vinegar. In the Philippines, where they are made, coconut and palm vinegar are frequently used. Middle Eastern dishes frequently employ vinegar derived from raisins. Despite the rarity of vinegar made from honey, it is commercially accessible in Italy, France, Romania, and Spain. Vinegar is made using cultures of Acetobacter aceti. With full air injection, these operate flawlessly at a temperature of roughly 28 °C. Typically, alcohol concentration should be less than 7.5 percent (v/v), and there shouldn't be any free sulfites.

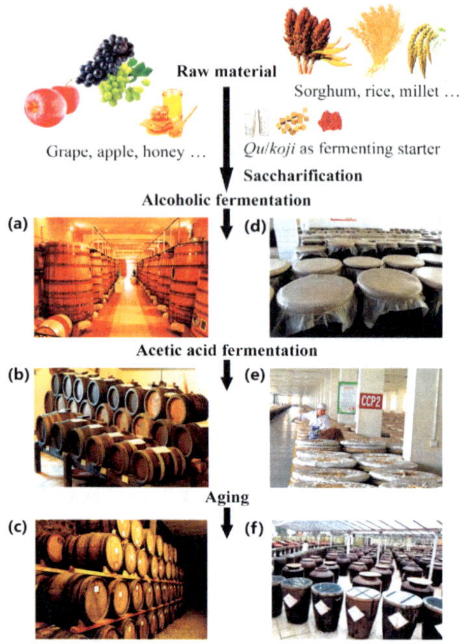

Figure 6.3. Vinegar.

(**Source:** *Zhang X-L, Zheng Y, Xia M-L, Wu Y-N, Liu X-J, Xie S-K, Wu Y-F, Wang M. Knowledge Domain and Emerging Trends in Vinegar Research: A Bibliometric Review of the Literature from WoSCC. Foods. 2020; 9(2):166. https://doi.org/10.3390/foods9020166, Creative Commons License)*

6.14 VINEGAR

Vinegar (Figure 6.3) mostly consists of acetic acid and only trace amounts of other molecules like alcohol, glycerol, esters, sugars, and salts. A chemical process that turns wine or fruit juice into vinegar allows for the production of acetaldehyde by partially oxidizing ethyl alcohol. Acetic acid is then produced from acetaldehyde. Vinegar involves two different types of biological processes: alcoholic fermentation and the oxidation of alcohol into acid. Alcoholic fermentation of carbohydrates is the first important stage in the making of vinegar. Yeast species are used to convert sugar to alcohol through fermentation. The second crucial step is Acetobacter species oxidizing alcohol to acid. Either a rapid or an extended procedure is used to make commercial vinegar. The fruit juices ferment into alcohol after a few days or perhaps months, and then they oxidize into acetic acid. Traditional vinegar is often made using slow techniques; the fermentation process takes weeks or months. The mother/starter of vinegar might accumulate and then be recovered. On the other hand, quick procedures boost oxygenation before adding air to the source liquid to achieve the fastest fermentation. As a result, vinegar can be made in a time frame of 1 to 3 days. The generator method and submerged fermentation method are used in contemporary production. These procedures are made to provide the alcohol product with as much oxygen as possible.

Orleans Technique is one of the most traditional ones for making vinegar that is still in use today(Webb & Staff, 2007). It was created in France and is defined as a continuous process. The starter culture is high-quality vinegar, to which wine is gradually added. In barrels with a capacity of 200 L, the vinegar ferments. 15 L of wine and about 70 L of high-grade vinegar are put into the barrel. A further 15 L of wine is added after a week. To replace the vinegar, more wine is put into the barrel after about a month. One of the main difficulties with this approach is how to introduce additional wine to the barrel without upsetting the bacterial mat on the surface. By inserting a glass tube into the bottom of the barrel, this might be avoided. Through the tube, more liquid is pumped in without disturbing the bacteria. Just below the level of the apertures on the ends, the barrel is filled. Insects can't enter the barrels since mesh has been put over the gaps. A 29°C ambient temperature is maintained. It is pumped out once the alcohol has turned into vinegar. There is still about 15% of the liquid in the barrel, which will be mixed with the subsequent batch.

Other techniques have been devised to speed up the Orleans process because it is slow. This technique makes use of a generator, which is a cylindrical tank and has a false bottom with holes in it. The generator is equipped with devices that allow an alcoholic solution to trickle through the splinters, where the acetic acid bacteria are injected. The mixture is a modified solution of alcohol with acetic acid and specific nutrients for acetic acid bacteria development. The mix is fluxed from the top of the chamber, allowing it to drizzle over the splinters before being collected at the generator's base. Up till the desired strength is achieved, the operation is repeated. Alcohol oxidation may cause temperatures to rise that are harmful to microorganisms. A cooling system is necessary to maintain the temperature below 25°C. With this approach, vinegar may be made in just 10 days, 1h is quicker than with other methods. The top of the container is filled with alcohol, which drips gradually. Air is blown through the perforations using an air compressor. From the container's bottom, vinegar is poured into storage tanks. This process yields vinegar with an extremely high acetic acid concentration. An oxygen-filled tank that contains alcohol is kept at a warm temperature. While nutrients and air are injected into the combination, the wine is maintained at a temperature of 26°C to 38°C. This process is frequently employed in the making of wine vinegar. Industry most frequently uses submerged vinegar systems to manufacture large amounts of vinegar with a high acetic acid content.

The liquid is continuously aerated for this system to function. Instead of producing vinegar, the bacteria that make vinegar are floating in the liquid. Since no slime develops in the machine, the vinegar looks clean. Turbine systems that are automatically regulated can be used to pump air into the liquid. These systems are made for manufacturers who intend to generate more than 50,000 L of vinegar annually. The Venturi is another often used air system. This is a less expensive and smaller system than a turbine. A closed stainless-steel tank is used for the processing, which is carried out via a pumping system. Air is introduced into the liquid using a Venturi air nozzle. This procedure takes around 30% longer than it would in a turbine system, but it is better for quality since there is less aeration, which prevents taste loss. The Venturi air system is currently the most popular processing method used by small and medium vinegar producers all over the world. Nutrients are pumped into the tank as the pump stirs the alcohol. Between 26°C and 38°C is the desired temperature range. Vinegar is made in a matter of hours. From the acetators, the vinegar is fed to a filtering device. To remove any sediment, the stainless-steel plates force the alcohol through the paper.

When manufacturing vinegar, two aspects need special consideration: oxygen supply and temperature. The mixture should be infused with oxygen, and the fermentation temperature should be maintained at about 28°C. Occasionally, unfavorable microbes could ruin the product. Turbatrixaceti (the vinegar nematode), for instance, may result in a decline in flavor and appearance. Therefore, it is important to check for contamination by harmful bacteria or molds. To prevent contamination, the generator must be periodically disinfected. Once the vinegar has finished fermenting, it needs to be kept in separate containers or wooden barrels. It is advised to find a location that is both dark and cold (under 10°C). The selection of appropriate starting cultures determines the final quality of vinegar. However, additional elements like the raw material, production technique, and aging procedure also have an impact on quality. There are some differences between goods produced on an industrial scale and those using traditional methods. Aromatic substances have a big impact on vinegar quality. The scent is a complicated fraction made up of several different chemicals with a wide variety of volatilities, polarity, and concentrations between a few milligrams and a few nanograms per liter. The scent of wine vinegar contains more than a hundred different chemical components, including volatile esters, acids, alcohols, phenols, ethers, lactones, and carbonyl compounds. When wine is aged, contact with wood results in an increase in aromatic complexity. In addition to their antioxidant properties, polyphenolic chemicals give vinegar its color and acidity. In the aerobic process of acetification, the bacteria's ability to multiply seems to depend on oxygen. Different kinds of polyphenolic chemicals are impacted by oxygen to varying degrees. For instance, during submerged fermentation, oxygen availability affects the flavonol content of vinegar. On the other hand, phenolic aldehydes that are emitted from hardwood barrels are unaffected by surface acetification vinegar. Reactions between compounds have a role in the aging process. Both polymerization and the release of substances from the wood are among them.

CHAPTER

7 Beverage Fermentation

Contents

7.1 INTRODUCTION

One of the most consumed foods in contemporary societies is fermented food and drink. To address the demand for high-quality processed foods and beverages that would be safer, devoid of synthetic chemicals, and at the same time nutritious and fresh, fermentation processes have developed over time. Novel processing techniques are also being investigated and put into practice. Many innovative methods have been used from the farm to plate/homes to suit consumer needs for fermented food items and to maximize consumer satisfaction. The modern food fermentation and beverage sector is very innovative and competitive, and it is constantly updating, innovating, and refining its technology for quality improvement and the production of new products from various food sources. The production and consumption of fermented foods are not steady over the world; it is important to highlight that the vast variety of fermented food products that are currently popular is attributable to global cultural diversity and local customer preferences. Dairy products, beer, wine, spirits, fermented fish, meat, and vegetables are some of the main categories of fermented food products. Since milk and milk products are used by six billion people globally and account for 6-7 percent of the dietary protein supply in Asia and Africa and 19 percent in Europe, dairy products have the biggest stake among the other categories of fermented foods (Oerke & Dehne, 1997). Given the probiotic microbial beneficial effects on health, demand for fermented dairy and non-dairy products is expanding quickly. One such innovation that has dramatically expanded the use of dairy products recently is the use of probiotics in dairy products like yogurt, cheese, ice creams, etc. To support human health, fermented foods have been strengthened with nutraceutical ingredients through improved fermentation techniques. Other examples include the use of encapsulation, genetic engineering, non-thermal processing for food safety and preservation, active packaging technologies, and computational intelligence methods for predicting problematic wines, beers, and other alcoholic beverages. These examples have all been regarded as cutting-edge systems that have significantly improved the food and beverage industries.

7.2 QUALITY

The products were unpredictable and of diverse quality since the principles and biochemistry of the microorganisms used in the indigenous and traditional fermentation processes were not understood. The understanding of the role of microorganisms in food fermentation and their modes of operation gradually encouraged food scientists to use some bacteria. Starter culture is a strain of microbe chosen for its stable properties to generate food with desired characteristics under regulated situations. Bacteria, yeasts, and molds are the three most prevalent types of microorganisms found in food fermentations. Single-strain, multi-strain, and mixed-strain are the three categories under which starter cultures are categorized. The Lactobacilli, which can produce lactic acid from carbohydrates, are important bacteria in food fermentation. The genera Acetobacter and Bacillus, which produce acetic acid mostly through the fermentation of fruits and vegetables and legumes, respectively, are other important bacteria. The Saccharomyces family, notably S. cerevisiae, has yeasts that promote excellent food fermentations. Because they produce enzymes that lead to beneficial biochemical reactions that contribute to the typical flavor and aroma in wine, beer, and ethanol as well as layering bread, yeasts play a significant role in the food sector. However, the bacteria that are frequently found in fermented foods are called lactic acid bacteria (LAB). Their importance is connected to physiological traits like substrate consumption, metabolic competence, and probiotic qualities. They are accepted as GRAS (generally regarded as safe) for human consumption because they are frequently found in foods and have a long history of use. From various fermented foods, different LABs have been isolated.

7.3 FERMENTED FOODS

For centuries, people have used fermented foods as a source of nutrition. Due to their potential health advantages, fermented foods and newly developed functional foods are becoming more and more popular. Recently, studies on these food potential medical advantages have focused on their antihypertensive, antidiarrheal, blood glucose-lowering, and antithrombotic qualities, to mention a few. In contrast to their nonfermented counterparts, fermented foods contain significantly higher levels of vitamins, minerals, and amino acids as well as bioactive compounds in the form of phytochemicals like phenolics, flavones, fatty acids, and saccharides. Foods that have been fermented are known to gather/accumulate bioactive chemicals during processing that are naturally lacking or only mildly present in their untreated counterparts. For instance, fermented ginseng roots have higher quantities of saponins than red ginseng roots do. Red ginseng roots include bioactive chemicals including saponins (ginsenosides) and non-saponins. These saponins are known to control the levels of insulin and blood sugar. One of the main foods consumed in Korea, China, Japan, Indonesia, and Vietnam is fermented soybeans, which have been shown to have anti-diabetic properties. When the fermented product is processed, small molecules undergo both quantitative and qualitative modifications that contribute to the characteristics of soybean. Certain foods and drinks that have undergone fermentation using particular LAB strains are known to have immune-regulating and non-allergenic effects. Recent advances in science have made metabolomics, microbial ecology, and genomes tools for determining the disease-fighting capabilities of bioactive chemicals and discovering novel ones from fermented foods.

A centuries-old method of food processing, fermentation has been passed down through the generations. Modern technology can ensure quick delivery of desired fermented foods while maintaining consistency in superior quality, sensory attributes, and nutritional benefits in light of the growing population, making it impossible for conventional fermentation techniques to meet consumer demands. With the emergence of new technology in food science research, novel food processes are being adopted in the food and beverage industries. Co-culture, thermophilic fermentation, molecular tools, genetic engineering, mutant selection, and recombinant DNA technologies, among other revolutionary innovations, can be effectively used to design and build custom starter cultures that outperform those found naturally. For instance, by combining starting cultures of Saccharomyces cerevisiae and Pediococcusacidilactici, S. cerevisiae and Lactobacillus acidophilus, P. acidilactici and L. acidophilus, and S. cerevisiae, P. acidilactici, and L. acidophilus, beverages have been produced.

7.4 THERMAL TREATMENT

The thermal treatments are meant to increase the shelf life, eliminate harmful bacteria, increase metabolic activity and enzyme synthesis, and accelerate the fermentation process of fermented foods (Nout, 1994). Radio-frequency heating and microwave heating are a few thermal processes. Thermal processing, however, alters the biochemical, textural, and sensory characteristics of fermented foods. High-pressure processing (HPP), high-pressure carbon dioxide (HPCD), ultrasound, ultraviolet radiation (UV), gamma radiation, microwave radiation, and pulsed electric field (PEF) are the most popular innovative nonthermal technologies utilized in the fermentation sector. The use of irradiation for food processing, which is derived from cobalt-60 or cesium-137 and UV radiation for surface sterilization, has been allowed by the FDA. According to the US FDA, UV radiation is a nonionizing ray that either kills or prevents the multiplication of bacteria by interfering with DNA replication by forming thymine dimers in the chromosome. To assure sterilization, HPP is applied to both solid and liquid food products in a range of 100-800 MPa at various temperatures. It eliminates the contaminating microorganism's cytoplasmic membrane. Cell death is primarily caused by solute loss, enzyme deactivation, and protein coagulation. HPCD is regarded as an energy-efficient and environmentally beneficial technique because it uses minimal pressure and harmless CO_2. Food is placed between two electrodes and sterilized using pulses of a strong electric field (1-50 kV/cm) for brief durations using PEF, which is suited to liquid and semi-liquid samples. Various natural antimicrobials are also added in the form of extracts from herbs, spices, and vegetables in addition to these nonthermal procedures.

Seaweed, chili, and green tea extracts stop Salmonella sp. growth. Carvacrol, citral, thymol, eugenol, and citric acid, among other substances found in the essential oils of herbs like rosemary, basil, ginger, and sumac, have been thoroughly investigated for their antibacterial action. Nonthermal procedures quicken chemical reactions occurring during processing, such as oxidation, polymerization, condensation, and esterification, as well as pasteurization, which increases fermentation rates and keeps track of the fermentation process. Other methods that have recently developed alongside thermal and nonthermal processes are nanotechnology and metabolomics. Even though fermented foods are packed with vitamins, minerals, and other beneficial ingredients, these may be lost during processing. Nutraceuticals have been successfully protected from deterioration using nanoemulsion forms, which also improve stability, bioavailability, and the safe distribution of useful components to potential consumers. Nano-sensors are used to properly monitor and assess the quality characteristics of fermented foods. Quantum dots, which are biosensors for identifying harmful and pathogenic microbes that cause food spoiling, are important. Although there are many alcoholic drinks consumed around the world, the majority of them are local and unknown in other regions. Beer, wine, and spirits including whiskey, vodka, gin, and rum are the most widely consumed alcoholic beverages worldwide. Breweries offer seasonally appropriate beers with a variety of flavors and alcohol content. Distillation units have increased in recent years due to their ability to produce novel aromas and flavors. Breweries and distilleries can produce different production batches using a variety of raw materials and different fermentation procedures. Modern wine production employs newer tropical substrates with unconventional organoleptic properties.

7.5 TROPICAL WINES

Mango, jackfruit, litchi, sapota, and other tropical fruits have been used to make tropical wines (Sarkar et al., 2022). In contrast to non-GM yeast strains, certain genetically modified (GM) yeast strains produce tasty flavors and exploit the residual sugars in beer to speed up fermentation. Similar to this, Saccharomyces cerevisiae ML01, a yeast used for wine fermentation, was genetically altered to reduce the formation of biogenic amines, hazardous by-products of wine fermentation. It is known that a different recombinant S. cerevisiae strain can lessen the production of the carcinogen ethyl carbamate, which is formed during the fermentation of wine. Fast molecular techniques like PCR, RAPD-PCR, PCR-TTGE, PCR-DGGE, short tandem repeats, etc. are used to characterize a specific starter culture and undesired yeast or bacteria. Even though GM yeasts with effective alcohol production and tolerance capability have been produced, customers' lack of interest in GMOs makes adoption problematic. As a result, very few reputable breweries or distilleries use GM yeast as a starter culture. Functional foods are processed foods with components that provide certain health benefits in addition to nourishment. Due to its therapeutic and preventative characteristics, the inclusion of probiotic starter cultures in the production of fermented dairy products is highly interesting as these bacteria increase the population of good intestinal microflora.

7.6 PROBIOTICS

The probiotic genera Lactobacillus and Bifidobacterium are important, while a few other species from the genera Pediococcus, Enterococcus, and Lactococcus are also of importance. However, selection should ideally be based on factors such as their GRAS (generally recognized as safe) status, continued safety in food applications, lack of pathogenicity, stability in acid, etc. A lot of technical parameters need to be optimized for probiotics to survive during the fermentation process and storage. In addition to giving foods texture, prebiotics in fermented food encourage the gut microbiota. Fructoolignosaccharide (FOS), inulin, and pectin were added to probiotic-fermented soymilk to raise the angiotensin-converting enzyme inhibitory activity, which improved the antihypertensive impact Varied starting cultures, such as bacteria, yeasts, and molds, offer a wide range of health prospects with novel bioactive components. The popular dairy products that are fermented are yogurt, cultured buttermilk, acidophilus milk, zabadi, labneh, biogarde, bifighurt, and dahi. Kefir and acidophilus-yeast milk are alcohol-lactic fermented dairy products. Prebiotics (Figure 7.1) and probiotics, which have distinct nutritional properties are extra functional elements present in fermented dairy products.

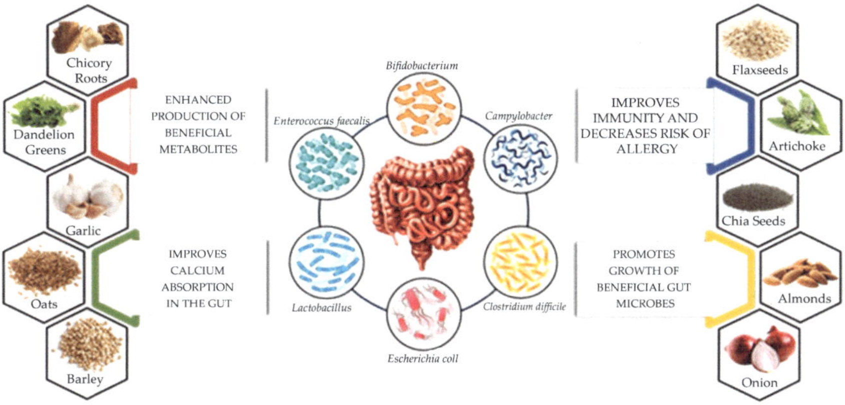

Figure 7.1. Prebiotics.

*(***Source:** *Kaur AP, Bhardwaj S, Dhanjal DS, Nepovimova E, Cruz-Martins N, Kuča K, Chopra C, Singh R, Kumar H, Şen F, Kumar V, Verma R, Kumar D. Plant Prebiotics and Their Role in the Amelioration of Diseases. Biomolecules. 2021; 11(3):440. https://doi.org/10.3390/biom11030440, Creative Commons License)*

7.7 ANTIOXIDANTS

Antioxidants derived from plants have made a significant contribution to shielding cells from oxidative stress. In vitro tests have been used to estimate their functionalities, but new developments in molecular biology have made it possible to anticipate the functional characteristics using omic techniques. Culture selection entails evaluating probiotic potential, effectiveness in reaching the colonization site, and safety while taking the consumer's health into account. Recent developments in molecular biology have made it possible to anticipate a wide range of characteristics, including antagonistic activity against pathogens, enzyme synthesis, colonization at the target site, and antibiotic resistance. Probiotic cultures have been bioengineered to give them the desired characteristics or to deliver specific bioactive molecules for the prevention and/or treatment of a variety of diseases, including antimicrobial activity, antitumor activity, and antidiabetic activity (Sleator & Hill, 2008). The success of fermented milk, yogurts, and fermented meats has made the health benefits of probiotics on humans extremely apparent. However, consumer preferences for vegetarianism, the consumption of beverages with strong flavors that are also extremely nutritious, such as fermented juices and smoothies, and the popularity of foods that resemble yogurt are growing. Food corporations have been investing in the formation of new functional and nutrition-modified foods due to market expansion and strong profit margins. Although there is a sizable market for dairy products, there is also a growing market for non-dairy beverages, which is anticipated to increase annually over the coming years. Fermented fruits and vegetables, among other non-dairy probiotic products, have already entered the market. When producing fermented vegetable juices, starter cultures are typically introduced to raw vegetables and mildly heat-treated vegetables to induce spontaneous fermentation by autochthonous microbiota or by starter cultures.

Bifidobacterium and lactic acid bacteria with low GC percent make up the majority of probiotic starter cultures. Yeast plays a crucial role in traditional fermented foods that have been linked to health advantages. The strain utilized for the starter culture as well as the substrate employed have an impact on the viability and functionality of the microorganism. It has been noted that orange and pineapple juices allow Lactobacillus sp. and Bifidobacterium sp. to survive longer. In addition to the human gut, probiotic bacterial strains can be isolated from fresh fruits and vegetables. These strains can serve as an alternative to novel probiotics, which can adhere to the gut epithelium and improve the health of the consumer. Adding yeast autolysate to the juices increases the number of lactic acid bacteria present during the fermentation process, shortening the fermentation period. Smoothies are among the best examples of fermented beverages made from vegetables. The majority of fermented beef products are produced in Europe. Their abundance of proteins, lipids, vital amino acids, minerals, and vitamins makes them a popular food. Research investigations have concentrated on the identification of genetically modified starter cultures and the function of enzymes in the development of the physical and sensory properties of fermented meats to produce more reliable, affordable, and consumer-friendly fermented meals. The manufacture of fermented sausages and cured meat is influenced by numerous internal and external elements, including starter cultures introduced to the product, the quality of the meat, fatty tissue, and additives, as well as salting, curing, casing, preparation of the sausage batter, and microclimate. Product quality and safety are both impacted by all of these conditions. The primary factor that has prompted the development of novel technologies that would impart stability and increased safety while keeping the intrinsic features of the products is, however, the safety of the final processed food. Consumers must prioritize safety to protect themselves from toxicological and microbiological risks. For decontamination purposes and process optimization, nonthermal processing technologies like high hydrostatic pressure (HHP), pulsed electric field (PEF), X-ray irradiation, pulsed ultraviolet (UV), and power ultrasound are being considered across a wide spectrum of meat products. These technologies may be useful for fermented meat processing.

Complex fermentation uses a defined or undefined starting culture to ferment a substrate into a product like cheese, malolactic wine, fermented seafood, etc. Studying the microbiome at various stages of fermentation is crucial for ensuring product safety and quality (Thomsen, 2005). rDNA amplicon sequencing was used to characterize the bacterial and fungal populations. The identification of fungal and bacterial operational taxonomic units revealed that fungal diversity decreases and bacterial diversity increases as a result of fermentation in both raw and ripened foods. A fermenter, also known as a bioreactor, is a device used to biologically process food and other bioproducts in a controlled setting. Fermenters are specifically designed with the end products and raw materials in mind. Bioreactors are divided into continuous, fed-batch, and batch forms of operation based on the feeding of the nutrients. The design of the bioreactor involves several crucial engineering components that are continuously updated and modernized to improve and enhance the quality and productivity of the finished product. The vessel design, height-to-diameter ratio, kind of agitation and agitator, aeration, and baffles are crucial components of a bioreactor. Self-cleaning micro spargers, mechanical foam breakers, temperature, pH, and pO_2 probes, valves, steam traps, and sample ports are other recent additions to the fermenters. The automation of fermenters or bioreactors through computer coupling is the most important invention. Computers play three key roles in food fermenters: (a) logging data through data acquisition, which consists of both hardware and software components; (b) data analysis, where data processing is carried out by a system based on a series of mathematical equations; and (c) process control, where signals from the computer are received and a program directs for precise operation of pumps, valves, and switches.

7.8 PROCESSED FOODS

Processed foods must be packaged to retain their quality, freshness, and safety. With changing lifestyles, packaging has improved from the initial simple paper and carton packaging to contemporary vacuum-processed packaging. Due to the contemporary challenges, the food packaging sector is constantly changing. Earlier forms of food packaging included metal, glass, paper, plastic, wood crates, etc but consumers choose microwave packaging, single-serve packs, etc. The food and beverage packaging businesses are advancing technically and lowering environmental stress caused by packaging waste and food loss while assuring a fresh, safe, and extended shelf life for foods and beverages. Modern food packaging has changed as a result of recent advancements in active packaging, including antimicrobials and antioxidants, nano-packaging, aseptic packaging, packaging mechanisms controlling volatile flavors and aromas, radio frequency identification (RFID) and electronic product codes, as well as active packaging involving oxygen, carbon dioxide, ethylene scavengers, and moisture control agents (Siro, 2012). The purpose of active packaging is to absorb the unwanted compounds that are released from the packaged food to prevent food deterioration and preserve freshness. Chemicals and biosensors are used in smart packaging to check quality and safety. As a result, an online quality control system with effective sensors built into the packaging should be made to double-check the product quality in terms of weight, volume, color, appearance, and biochemical composition. There is a big market for bio-formulations based on food packaging as a result of rising environmental pollution concerns.

One of the main goals of technological advancements and innovations in the food sector has been to satisfy both the desire for such unique and healthy foods on the part of customers as well as their acceptance of these foods. Some technological improvements have been readily accepted, while others, like food irradiation and genetically modified foods (GMFs) in Europe, have been more easily rejected. Therefore, the acceptability of consumers plays a key role in the long-term success of a new food product. Innovation generally entails turning a commercially viable idea into a successful product. This process comprises a wide range of activities, including marketing, technology, partnership, commercialization, safety concerns, and risk factors. The adoption of innovative foods by consumers is influenced by a variety of elements, including their participation in product processing, cultural, social, and lifestyle considerations, long-term implications of novel technology, such as hazards and concerns for human health, etc. Due to technological advancements, customers are becoming more informed, and they are routinely evaluating the possible health benefits of functional foods and dietary supplements. Consumers are skeptical about the long-term health hazards of these functional foods, although scientific studies have shown them to be safe. A substantial portion of the public is looking for minimally processed, clean, and green foods and organic foods.

7.9 GENETIC ENGINEERING

Incorporating or altering a targeted gene from an organism which could be a plant, animal, or microorganism leads to the production of genetically engineered foods. Three methods are typically used to modify genes: (a) direct DNA transfer using microparticle bombardment or gene knockout or gene addition; (b) DNA transfer using bacterial vehicles like T-DNA; and (c) editing of genomic DNA. The fermentation industries now have improved product yields, superior quality goods, and the ability to produce unique foods. For instance, the use of transcriptomics methods has helped to explain the molecular mechanisms behind the metabolic changes that fermenting bacteria undergo. Recently, the entire genome profiles of fermenting yeasts and bacteria have been made available in public databases. Genomic databases have made it possible to examine gene expressions under various situations, giving fresh insight into metabolic processes. Microarrays were used to study the stress factors in the fermentation process affecting the transcriptional response for laboratory or industrial wine, lager brewing, and yeast strains, as well as gene expression dynamics during the various stages of fermentation of synthetic media or natural substrates (Ye et al., 2001). In addition to yeasts, LAB also has industrial value since they may give a variety of fermented foods, like dairy products, and fermented sausages, their distinctive flavors, appealing textures, and preservation characteristics. To analyze the gene groups involved in metabolic and functional processes, gene expression microarray has proven to be quite beneficial. The ability to investigate novel elements in the transcriptomes of fermenting microbes has improved due to RNA-seq. For food science and the industry, it is crucial to profile microbial communities and elucidate the molecular mechanisms behind the metabolic processes and pathways present in fermented food ecosystems. The novel omics method known as metatranscriptomics has significantly advanced our understanding of microbial behaviors in food ecosystems and enables us to use them to increase food and beverage production and quality.

7.10 GENETICALLY MODIFIED ORGANISMS

Genetically modified (GM) microorganisms are created to either enhance a desired product or prevent the production of unwanted metabolites. Trehalose is famous for accumulating 30% of yeast cells under temperature stress, which is a serious issue for the baking sector. A novel yeast strain with tolerance to stress, including dehydration, freezing, and hazardous doses of ethanol, was created by interrupting the trehalose (ATH_1) gene to solve the aforementioned issue. Domesticated, Saccharomyces and S. eubayanus strains were combined to create yeast hybrids, which produced more ethanol (5.6%) and fermented food more quickly than their parent strains (4.5%). A recombinant S. cerevisiae strain was created that has high glucoamylase activity and accelerates maturation. However, the new yeast does not contain any heterologous DNA sequences, bacterial sequences, or drug-resistant genes, making it safe to use in breweries. New yeast strains with the ability to produce high-flavor chemicals like terpenoids must be created by incorporating beta-glucosidase genes from the Aspergillus, Hanseniaspora, or Pichia families into wine yeasts. Important B-group vitamin folate is known to guard against cancer and heart defects. A small number of LABs are known to produce folic acid.

Nevertheless, despite many benefits, GM foods come with several health and ecological risks, such as competing with wild species, which have put consumers in a difficult position. Consuming transgenic soybeans and maize has been linked to allergic responses. Regarding GM foods, different countries have varied laws. Most nations have laws and regulations that permit manufacture and commercialization. However, other nations, including Israel, Norway, the Netherlands, and Russia, forbid the manufacturing and sale of GM foods within their borders. Even though GM foods are permitted in China, France, South Korea, and New Zealand, the transition is taking a very long time. China has not put any explicit GM food laws into effect. In China, it is legal to cultivate and spread GM papaya. Similar to other countries, the USA has limited regulations and no federal legislation that specifically addresses GM foods. While the USA and Canada have relatively lax regulations on the use of GMOs in food, the European Union (EU) has highly strict regulations. The EU accepts genetically modified (GM) microorganisms resulting from uncontrolled genetic changes like random mutation, dominant selection, adaptive evolution, natural conjugation, and transformation, but recombinant DNA technology-derived microbes are typically disallowed for use in the food and beverage industries.

Natural resources and the supply of food are being depleted as a result of population growth and environmental changes. This has made it necessary to conduct scientific research on the current methods and replace them with novel ones to increase the production of meals that may be beneficial for your health. To protect innovations and ideas from infringement, intellectual property rights (IPR) have become increasingly important in the food and beverage industry due to the fast-changing needs of consumers. Patents, copyrights, trademarks, industrial designs, trade secrets, and geographical indications are just a few of the several IPR protections available in the food and beverage business. Our diet consists mostly of fermented foods, which have been consumed for centuries. Fermented food preparation in small-scale or household settings requires only a few basic techniques and tools. An older and more affordable method of generating and protecting food against food deterioration is fermentation. In addition to preservation, other benefits include flavor enrichment, enhanced digestibility, and increased nutritional and therapeutic qualities. By the mid-19th century, the industrialization of fermented foods and the discovery of microorganisms had a significant impact on the growth of both fermented foods and fermentation methods utilizing cultures. Interestingly, milk, bread, vegetables, and meat are just a few of the commercially relevant foods where yeast and lactic acid bacteria (LAB) are the principal players contributing towards fermentation processes. Even if the production of fermented foods has increased, traditional fermentation is still used in some nations because of its superior flavor and aroma qualities. However, it seems inevitable and even strange when the demand for these foods increases and their consumption becomes widespread; the solution to grow the market may be achieved by scaling up the industrial process, in which the utilization of starter cultures is a major factor.

7.11 SCREENING

Currently, strategy rather than screening is used to develop starter cultures for fermented foods. The foundation of the strategy is a knowledge of organism physiology and metabolism, as well as their importance in food items. Innovations in genomics will make it easier for us to build solid strategies by giving us reliable data. The technologies for designing food-grade products include high-throughput selection methods, laboratory automation, genomics, proteomics, and metabolomics. Future regulation patterns will be greatly improved by the evolution of new data. It is currently very difficult to make predictions about how it will affect the future of regulatory requirements in the food business. Applications of biotechnology may support the production of healthy products or may limit the use of a few strains as starter cultures with valid authorization. The establishment of starter cultures in the sectors of biosafety, probiotics, and the potential use of cultures in other fermentable foods are the results of recent breakthroughs. In anaerobic conditions, microorganisms like bacteria, fungi, or yeast produce fermentation, a biotechnological process that has been employed by humans for at least 6,000 years. Fermentable carbohydrates are transformed into products like alcohol, organic acids, and carbon dioxide during the process. The primary objective of its role was to preserve food and increase shelf life, while also ensuring food safety, nutritional value, and organoleptic quality.

Both high and low-income individuals continue to consume a lot of fermented foods. Every community adheres to its unique traditions about fermented foods, which represent the traditions, agro-economic, and sociological characteristics of that community. Some fermented foods are popular as delicious regular dishes and are promoted globally for their functional, nutritional, and medicinal values; however, fast foods with high calorific values are replacing traditional food habits in Asia and Africa due to modern civilization, which reduces the usefulness of traditional fermented foods. Foods that are fermented have a sizable global market and are one of the most actively researched areas of food science. The rise in popularity of fermented foods is attributed to consumer well-being, improved healthcare, and increased knowledge of the many advantages of functional foods, which encouraged their industrial manufacturing. The disease prevention and disease resistance of fermented foods are only a few of their many benefits. Currently, a wide range of fermented products promoted in both industrialized and developing countries highlights the importance of fermented foods for consumers. These products are valued for their guaranteed benefits of safety and preservation as well as their highly palatable sensory qualities, which are supported by microorganisms and their enzymes. Due to the elevated biological chemicals found in fermented foods, which are produced by oxidation, reduction, hydrolysis, condensation, and isomerization, they are recognized to have a positive impact on human health. Regular consumption of fermented foods boosts beneficial bacteria in the gut by up to 10^3-fold, improving the health of the GI tract. In the past, inoculum from a prior production was utilized as a starter for fermentation. A starter culture is a combination of useful microorganisms that typically includes a cultivation medium that is populated by bacteria involved in fermentation. With the discovery of new microorganisms, the production of fermented foods was increased using starter cultures that had been thoroughly characterized. This sped up the fermentation process and allowed for the production of fermented foods with various sensory and dietary properties.

In starter cultures, bacteria, yeasts, or molds are present. Yeasts and LAB are important in the process of fermentation (Bevilacqua et al., 2013). Faster acidification and desired aroma, consistency, taste, texture and safety attributes are advantageous features of starter cultures. In Denmark and Germany, starter cultures were first isolated and produced in 1890 from cheese and sour milk. However, modern scientific techniques have developed functional starting cultures, which provide at least one functional feature to improve food product safety, quality, or nutrition. To increase the use of starter cultures in fermentation, first, starter cultures with certain traits were extracted. Random mutagenesis and selection methods were applied. Normal microbial techniques are still in use, although genetically modified microorganisms received less acceptance. New DNA sequencing techniques, however, make it easier to quickly look for advantageous random mutations in the genome. These techniques require new tools for analyzing and combining the data gathered; bioinformatics plays a major role in predicting the favorable and unfavorable features of microorganisms, while laboratory tests are required to verify their growth in various food mediums. Although starter cultures are chosen for safety and qualities that may benefit consumers, they must be generated under industrial conditions. It is essential to research the function of the culture and create strategies for enhancing the function of the culture to produce the optimum culture for any food product. Up until recently, the search for a starting culture has relied on the screening of various isolates in small-scale food fermentations. Based on their performance during the process and a successful organoleptic study of the food product, starting cultures are chosen. With the help of this technology, enormous cultures have been separated, and in the future, it will also be used to increase the microbial collection to be utilized as starting cultures. Modern techniques have made it possible in recent years to precisely target certain genes and metabolic pathways that are responsible for key starter culture operating factors.

Regular starter cultures can be classified as mesophilic or thermophilic based on the cultivation temperatures used for their growth and production. Although thermophilic cultures work best at a higher temperature (42°C), mesophilic cultures typically grow and release lactic acid at a mild temperature (30°C). Lactococcus lactis subsp. lactis, L. lactis subsp. cremoris, and Leuconostocmesenteroides subsp. cremoris are examples of mesophilic starters, whereas Lactobacillus delbrueckii, L. helveticus, and Streptococcus thermophilus are thermophilic starters. Whole starting cultures that are in use today were developed in one way or another from natural starters that were a combination of different strains and/or species. Industrial mixed-strain starters (MSS) have replaced natural starters for a few products (Feutry et al., 2016). Traditional starters contain a variety of microbial strains, including yeast, bacteria, and molds, which affect the complexity of the final product. Starters that are free of undesirable organisms are necessary for commercial production. Such goals are difficult to reach using conventional techniques. The growth of natural starters dates back to the ancient practice of back slopping, which involves inoculating a new batch with a previous batch of fermented material and/or subjecting it to specific environmental factors like high heat, low pH, or incubation temperature.

7.12 PRECAUTIONS

No special safety precautions are needed during starter cultivation to prevent contamination of the media and culture settings as well as from external sources. Natural starters are a valuable source of strains with the right technological traits, like aroma and antibacterial production together with phage resistance. Starter companies and research organizations store, cultivate, and supply MSS, which are obtained through a careful selection of natural starters. Similar to natural starters, MSS contain an unspecified mixture of strains with varying physiological and technological traits. Undefined strains are grown in controlled circumstances with few subcultures, yet in comparison to natural strains, the hardness of their composition and function is greatly improved with less internal variability. The starter tank must be inoculated using concentrated cultures instead of the small amounts of stock cultures used in traditional MSS cultivation, which necessitates many shifts inside the factory and increases the risk of instabilities in starting components and function. Defined Strain Starters (DSS) are preserved, cultivated, and supplied by specialized industries. DSS is made up of one or more cultures. DSS currently substitutes traditional starters. Phage contamination could prevent lactic acid fermentation due to the usage of a few strains. It is difficult to preserve the unique traits due to the failure of natural microbial diversity. However, evaluation of the essential aspects of each species, such as growth and acid production, and genetic or biochemical factors, might result in the reasonable blending of strains, allowing for the formation of cultures with the right properties. A specific strain or diverse population of microorganisms is multiplied in starter culture production to reach a concentration that is anticipated to remain and be metabolically active throughout the process. Production of industrial starting cultures follows a specific procedure and includes necessary quality controls.

The safety and adaptability of food production may be impacted by the choice of starting cultures made on-site or purchased from commercial providers. Some food sectors provide their starting cultures to other businesses. The use of freeze-dried starting cultures prevents internal culture reproduction, which lowers costs associated with culture manufacturing and lowers the risk of bacteriophage contamination. The manufacturing of a stock culture, culture media preparation, growth of the stock culture to the desired cell density in a bioreactor, collection and concentration of cells from medium, and culture maintenance are the key stages in the industrial production of starting cultures. The development of the desired purity and quality of the culture product depends on each step in the production process of starter cultures. Industrial starting cultures are often provided as dehydrated or frozen microbial cells and are produced by some companies. To prevent loss during storage and distribution, these procedures call for culture concentration and dehydration in addition to preservation.

Modern manufacturing techniques for industrial starting cultures require process engineering, as well as multidisciplinary knowledge of microbiology and microbial physiology. Expertise in cryobiology is required when using cryotechnologies to preserve cultural traits. The evaluation of foods, maintenance and management of plant cleanliness, and examination of raw materials are only a few of the many procedures that make up overall quality control. It's critical to sustain inoculum quality, and hygiene throughout the entire manufacturing process. The conditions for starter culture formation must be repeatable to produce an end product with consistent quality. As a result, tests for cell viability and contaminant detection are used to evaluate the quality of batches. The HACCP system also affects the quality of the fermentation process and is followed while manufacturing starter cultures for industrial use. Starter cultures must be produced at each stage in conditions that support cell survival, maintain cell viability during storage, and verify the cells' morphology/condition in fermentation. The development and survival of starting cultures may be impacted by exposing them to strain at different production stages. Therefore, it is essential to identify and concentrate on stress-stimulating variables at every step.

LAB are the most common microorganisms used in the food industry. LAB converts the carbon source present in raw foods to lactic acid while simultaneously reducing the pH, which has a substantial impact on the removal of undesirable organisms, enhancement of sensory and textural features, as well as impact on health benefits. LAB increases the nutritional value, flavor, and texture of fermented foods, including dairy products like buttermilk, cheese, yogurt and fermented milk, fermented beverages, fermented meats and vegetables. LAB are recognized for producing probiotic-like enzymes, sugar polymers, vitamins, antibacterial and aromatic chemicals, and sugar polymers. Functional characteristics can also be related to GI tract survival and maintenance, viability in low and high-pH environments, bile salt tolerance, antibiotic resistance, hydrophobic properties, immunomodulation, and antagonistic and antimutagenic properties. Technological factors include the ability to grow at various pH, temperature, and salt (NaCl) levels, as well as their capacity for acidification and metabolism i.e., deamination of arginine, hydrolysis of aesculin, production of acetoin, all of which tend to produce adequate flavor and texture. Regarding the salting effect, it is a common practice to supplement NaCl in the majority of fermented dairy foods, which also has an impact on the starting culture growth. The primary role of starter cultures in the synthesis of milk is peptide breakdown, which allows the coagulant to produce tiny peptides and amino acids. Starter cultures break down caseins and transform amino acids into a range of flavoring substances.

The starter cultures release the intracellular proteolytic enzymes like peptidases and amino acid-degrading enzymes, into the cheese matrix through cell lysis for flavor development as well as to control bitterness in ripening cheese. Amylolytic LAB (ALAB) is an efficient method for completely converting starch into lactic acid by combining fermentation and saccharification in one step(Reddy et al., 2008). In addition to probiotics or hypoallergenic children's diets, ALAB strains may be effective during the fermentation of cereal-based foods and beverages. Extracellular amylases that produce ALAB are utilized as starters in dough technology to enhance bread texture and product shelf life. During glycolysis, lipolysis, and proteolysis reactions, cheese-ripening methods have used proteases and peptidases from LAB to acquire a variety of sensory qualities. Casein and whey hydrolysate are produced in the dairy industry using noncoagulated proteases as an alternative. B-group vitamins, such as folate (vitamin B_9), riboflavin (vitamin B_2), cobalamin (vitamin B_{12}), and menaquinone (vitamin K_2), can be synthesized by several gastrointestinal LAB species. These vitamins can be found in LAB-fermented foods such as cheese, yogurt, and cultured buttermilk. Folate content is increased during fermentation using Bifidobacterium strains and starters like Lb. helveticus MTCC 5463 and Lb. rhamnosus MTCC 5462 in milk, Streptococcus thermophilus and Lb. delbrueckii subsp. bulgaricus in yogurt, and Streptococcus. thermophilus and Lb. delbrueckii subsp. bulgaricus in skim milk. GABA-producing LAB may use cheese, dairy products, black raspberry juice, kimchi, and soymilk as starters in the production of GABA-enhanced functional foods and beverages. GABA can be used as an ingredient in food; the US FDA has granted permission for the production of Pharma GABA™, a commercial natural GABA formulation made from Lb. hilgardii K-3. Because of the potential uses of EPS from LAB as a natural food additive for enhancing rheological characteristics, smoothness, stability, texture, and water retention in food items, EPS from LAB has higher economic value. The hardness, stickiness, volume, and mouthfeel of cereal and pseudocereal doughs were improved by EPS from Weissella spp. and Lb. sanfranciscensis starters.

7.13 PRODUCTION

For the production of fermented foods in the past, people mistakenly used yeasts; however, Louis Pasteur revealed in the 1860s that these microorganisms are capable of converting carbohydrates into ethanol and carbon dioxide (CO_2). In most bakery and pastry goods, yeast is utilized as a leavening ingredient to elevate the product. The yeast interacts positively with related bacteria in fermented foods by favoring the starter culture throughout the production of the final product or by avoiding and eliminating the unfavorable microorganisms that reduce the quality of the product. Through the production of antibiotic agents, organic acids, toxins, and hydrogen peroxide, yeast inhibits the growth of pathogenic organisms in fermented foods. By converting complex carbohydrates into vitamins, amino acids, and purines as well as free sugars, yeast increases the number of lactic acid bacteria participating in food fermentation.

CHAPTER

8

Advances in Food Fermentation Technology

Contents

8.1 INTRODUCTION

To create novel fermented food items, alternative complementary technologies have been used to various degrees of commercialization. Novel technologies can help food processors meet business demands for energy-efficient procedures as well as customer desires for safer and higher-quality goods. The modern fermentation sector is very competitive, and it has taken the lead in evaluating the possibilities of new technologies to enhance fermentation processes and produce better-quality foods. To increase the production and quality of fermented products as well as produce new products from a variety of sources, the food fermentation sector needs unique procedures. From Neolithic China through the early Roman era, there has been a link between fermented food and health, and the oldest evidence implies that fermentation was a key component of ancient civilizations. Asian-developed fermented foods, such as pickles, vinegar, sauerkraut, yogurt, cheeses, and a variety of fermented milk and traditional alcoholic beverage products, are still consumed around the world. Fermented foods like beer and wine were extensively employed in Asian diets and were also used as medicines. Despite a long tradition of preparing and eating fermented foods, people were not aware of the function of microorganisms, microbial enzymes, and their interaction with one another during the fermentation process.

8.2 LEGUMES

Africa produces fermented starch crops and food items made from legumes (Campbell-Platt, 1994). A variety of fermented foods, including bread, cheese, and yogurt, have been produced all over the world due to advancements in fermentation techniques over the years. Around 5000 different types of fermented foods and beverages are consumed globally, and fermentation is frequently used in the food and functional food industries. Products made from meat, dairy, soy, fruits, cereals, vegetables, and fish are examples of fermented foods. Due to their useful biological activities, which range from the prevention of chronic diseases like diabetes and cardiovascular disease to cancer prevention, the secondary metabolites produced during fermentation processes, which include antibiotics and peptides, are also known as bioactive compounds. In addition to enhancing nutritional content, bioactive chemicals produced during the fermentation process also extend shelf life while enhancing safety. In the last two decades, the commercial production of functional foods, medicines, biofuels, and biochemicals has placed an increasing emphasis on the preservation of structural characteristics and bioconversion capabilities of microbial cell factories during long-term storage. Today, efficient and cost-competitive bio-based processes mostly depend on reliable upstream cell propagation strategies. The degree of cellular stability attained during the post-cultivation and storage stages has an impact on both cellular propagation and future scale-up procedures because microbial preservation is the first step in any upstream bio-production technique.

8.3 PRESERVATION

Cell preservation is one of the upstream processes that are essential to ensure complete and effective microbial cell proliferation while sustaining metabolite yields, and productivity throughout any scale-up process (Pan & Sun, 2011). Improper and ineffective cell preservation techniques may lead to productive degeneration. Thus, to ensure the highest level of cellular stability during long-term microbial storage, various preservation approaches have been developed in response to the loss of cellular functionality during seed propagation. Unpredictable metabolic reactions and decreased fermentation yields are probably caused by instability and cellular fragility. To prevent microbial productive degeneration throughout the fermentation process, various preservation strategies have emerged over the past ten years. One of the biggest obstacles to the development and deployment of effective probiotic systems for functional meals is cell preservation. Industrial starter cultures, which include lactic acid bacteria, are particularly susceptible to stressful situations, and their long-term stability is severely jeopardized during production and storage. Industrial processes like freeze-drying and spray-drying can harm the survival of microorganisms. Even though several efforts have been made to promote the stability of microorganisms during various downstream processing steps, increasing the survival rates remains one of the key difficulties in the manufacture of industrial starters and probiotics.

The most widely used methods for long-term microbial cell preservation are drying and cryopreservation. To pursue an effective cell preservation strategy, it is essential to strike a balance between stabilization and cell damage. Preservation strategies have negative effects on the viability, stability, and functionality of microorganisms. Spray-drying has been chosen for microbial dehydration although freeze-drying is frequently used in long-term storage because of its processing flexibility and cost-effectiveness in industrial settings. Even while cryopreservation techniques result in the loss of viability, depending on the processing variables used, both cryo-tolerance and microbial culture functionality can be increased. Research has been focused on the interactions between variables including the use of cooling rates and cryoprotective chemicals over the past ten years to increase cell viability and long-term stability. Researchers have also been able to quantify physiological cell responses at the single-cell level while comprehending the effects of processing procedures on cellular robustness using sophisticated analytical tools. Since the degree of stabilization of the microbial cultures is strongly proportional to the moisture content, effective drying techniques are essential for achieving long-term storage stability. Historically, cryopreservation or freeze-drying has been used to maintain microbial cells for an extended period. Cryopreservation is the method of choice for preserving microbial cultures.

8.4 CRYOPRESERVATION

Even in the presence of potent inhibitors, freeze-drying can cause the loss of microbial viability and stability when delicate microorganisms like lactic acid bacteria (LAB) are involved. The storage survival is impacted by the cryoprotectant. Standardized freeze-drying procedures can provide freeze-dried goods with repeatable viability over long-term storage for fragile or resistant bacteria like Campylobacter, Aeromonas, and Vibrio. The properties of the cell surface must also be considered during a freeze-drying process in addition to the freezing media.

To prepare an intermediate product for a subsequent freeze-drying process or as part of an end product during downstream processes, the freezing rate is also important. Despite the minimal likelihood of intracellular ice formation, osmotic-driven water migration can increase intracellular solute concentration, which can be harmful to cells when freezing rates are high. From an economic standpoint, higher freezing rates are typically preferable. If the freezing rates are high, cellular damage brought on by the mechanical forces produced by ice crystals may also occur. Innovative microbial preservation techniques have also been described for use in culture collections. These techniques involve immediate drying steps for long-term storage without the need for freezing procedures. Spray drying is an effective dehydration method for the preservation of microbial cultures on an industrial scale. High temperatures must be used to speed up the process of water evaporation and ensure adequate storage stability by achieving a final moisture content of between 4 and 7 percent.

Stresses created during this process are a significant disadvantage in the application of this method for the large-scale manufacture of microbial powders, although scalable production has been established. Spray drying is acknowledged as a cost-effective process in the production of probiotics, with approximately ten times lower operational costs than freeze-drying, even though its viability depends on the strain. Recent technical advancements, such as low heat treatments, have highlighted the adaptability of spray drying techniques in the formation of starter cultures and probiotics. The success of this technology depends on the incorporation of protective carrier matrices to protect against high drying temperatures in terms of cell viability. Fluid bed drying specifically entails the evaporation of water from wet powder or particles using heated air. Due to its high rates of success, cryopreservation continues to be the primary long-term cell preservation technique. Without phenotypic or genetic changes, microorganisms can be cryopreserved at low or ultralow temperatures while retaining cell viability. Large ice crystals that develop at freezing temperatures have a detrimental effect on cell viability and can mechanically damage cell membranes. Therefore, ice crystal management is crucial for raising survival rates. The temperature, freezing rate, and freezing time are key operating parameters for preserving biological activity during storage. Probiotics retain more acidity at lower temperatures and for shorter periods. Utilizing a high freezing rate and a low storage temperature can increase cell resistance to freezing. The outcome of the freezing process is determined by the freezing rate because mechanical damage may result from the presence of ice crystals inside or outside of the microbial cells. Repeated subculturing, however, may result in changes to the flocculation process and time-dependent genetic instability. It's interesting to note that a cutting-edge freezing technique called Cell Alive System (CAS) has recently been created for the quick preservation of environmental samples (Prakash et al., 2020). A supercooled liquid phase is produced during the freezing process by using an alternating magnetic field, resulting in a uniform freezing process with little crystal formation.

The basic objective of any strategy used to preserve microbial cultures is to guarantee better cell viability following the subsequent procedures. By including protective elements like cryoprotectants or antioxidants in the media, cell viability following cryopreservation or dehydration processes can be significantly increased. Cryoprotectants are necessary to lessen cell damage from freezing if the storage temperature is below freezing. Although viability loss is unavoidable, there are now unique commercial treatments and techniques available to lessen the effects of the freezing process and subsequent long-term preservation. When freezing microbiological suspensions, cryoprotectants also referred to as lyoprotectants in freeze-drying processes are added to reduce the harmful effects of ice crystal formation and lower the freezing point. DMSO, polysaccharides, glycerol, and non-permeable additives are currently employed in microbial cultures as cryoprotectants. These cryoprotective chemicals adhere to the microorganism surface, coat their cells, and so offer protection from the growth of ice crystals during long-term storage. Skim milk is commonly used as a drying medium in freeze-drying cell preservation techniques because it stabilizes the components of cell membranes by covering cells with a protective layer. As cryoprotectants, protein and carbohydrate-based matrices have also been used. However, the strain affects the ratio of protection. More protective effects can be achieved by combining cryoprotectants. Osmotic stress and the development of extracellular ice can both be avoided by using penetrating agents. When antioxidants like sodium glutamate and betaine are present in the protective media, low ice crystal formation is attained during freezing. By enhancing the mobility of the acyl chains of fatty acids in the membrane core region, amino acids can also function as cryoprotectants in lactobacilli.

8.5 SUGARS

Sugars can build structures that slow down molecular interactions inside the cytoplasm in addition to their capacity to replace water. Sugars help preserve microbial cells by replenishing lost water in the membrane and avoiding protein aggregation by forming hydrogen bonds with polar groups. The impact on the survival rate of adding various combinations of carriers and thermal protectants depends on the strain. After a drying process, cell viability loss can be minimized by combining skim milk with carbohydrates like sucrose or lactose in a synergistic manner. Skim milk plays a key role in the formation of a protective layer on the cell wall that leads to the stabilization of the components of the membrane. Antioxidants like monosodium glutamate can be added to the carrier medium to increase the protective impact of polyhydroxylated substances like trehalose. In addition to maintaining membrane integrity and fluidity, these synergistic combinations may also help to preserve the enzymatic activity of important metabolic enzymes. During the freeze-drying of yeast cells, the composition of the drying matrices is also crucial. During freeze-drying operations, the ability of disaccharides to form hydrogen bonds is crucial for maintaining membrane integrity and protein structures. A variable number of galactose units are connected to two to eight glucose monomeric units to form galacto-oligosaccharides (GOS), which are polyhydroxylated carbohydrate-based molecules. In addition to serving as effective cryoprotectants, GOS has recently attracted commercial interest.

Proper cell preservation is not guaranteed when polysaccharides with high transition temperatures are used. Because of the high transition temperatures of GOS, they have a strong protective potential. The synergistic interaction of two prebiotic substances i.e., GOS and lactulose, might enhance lactobacilli protection against freezing processes. Due to the high viscosity and low mobility conditions produced by the combination, glossy matrices are formed, which limit molecular interactions. GOS combinations with a high amount of tri and tetra-saccharides can act as membrane protectors post-rehydration in addition to their protective ability throughout the dehydration process. Dehydration causes GOS to replace water molecules in the membrane structure and establish hydrogen bonds with the polar groups of the phospholipids and proteins. Prebiotics like inulin and fructo-oligosaccharides (FOS) have been recommended as agents that will preserve lactic acid bacteria during freeze-drying in addition to GOS. Cellular immobilization has been suggested as an effective substitute to improve the microorganism stability throughout the cultivation phase. Microencapsulation has emerged as a viable method for protecting cells while enhancing their stability during storage. This is in contrast to freezing and freeze-drying, which may result in irreversible protein denaturation and membrane damage with detrimental effects on cell viability. The type and concentration of the entrapment substance used, along with the procedure, determine how efficient encapsulation as a cell preservation strategy is. Size, porosity, and texture are factors that influence the degree of protection provided to the microbe. The composition of the covering material can also help to protect cells and improve the efficiency of the encapsulation process.

Another approach for encapsulating probiotic strains that protects them from freezing temperatures is microencapsulation in calcium alginate. Encapsulation exhibited a protective impact on the viability of probiotic strains when stored at 80 °C in the absence of cryoprotectants. However, encapsulation in alginate was unable to protect the probiotic cells that were enclosed at 20 °C because physical changes, such as large particle size and loss of spherical form, were discovered within a short time. Another practical method to enhance the cell viability of probiotic strains like Enterococcus during drying procedures, storage, and gastrointestinal transit is microencapsulation. The immobilized beads are typically dried via freeze-drying since the dehydration of the beads is necessary for the long-term preservation of the entrapped microorganisms. Dehydration by freeze-drying involves oxidative stress that could cause an osmotic shock to the microbial cells. Cryoprotective chemicals are incorporated into the entrapment media to promote microbial cell preservation during freeze-drying (Etemadian et al., 2011). Therefore, the protective impact of microencapsulation can be enhanced by trehalose and sucrose or lactose and trehalose addition to the entrapment media. It has been discovered that adding sodium ascorbate as an antioxidant and trehalose as a cryoprotectant to the formulation increases bifidobacteria survival after freeze-drying. To replace the water lost during the dehydration process and to prevent mechanical contraction and osmotic shock during freeze-drying, these chemicals work synergistically.

8.6 IMMOBILIZATION

Immobilization utilizing dry biopolymers has recently emerged as a successful strategy to maintain cell viability and functionality during long-term storage in addition to the usual microencapsulation methodologies. To maintain the viability of delicate microorganisms, high-voltage electrohydrodynamic methods like electrospinning and electrospraying have recently become revolutionary microencapsulation techniques. Due to their higher surface area to volume ratios and high permeability, electrospinning and electrospraying have recently become effective entrapping methodologies for preserving microorganisms in addition to their promising applications in the food industry. The thin polymeric fibrous material produced in electrospinning methods specifically enables the entrapped cells to interchange nutrients and metabolic products while maintaining their metabolic activity. Both electrohydrodynamic techniques can entrap cells using a variety of support matrices, including polysaccharides like chitosan, cellulose, and alginate as well as protein-based substances. Electrohydrodynamic procedures also have the benefit of not requiring temperature control, allowing for the avoidance of negative effects on microbial physiology brought on by high processing temperatures. Electrospinning has recently been used to successfully entrap bifidobacterial strains by creating fibers that allowed for the maintenance of high microbial viability despite the osmotic change and electrostatic field produced during the encapsulation process.

The integration of microbial cells into nanofibers using electrospinning gives benefits in terms of protein stability and functionality that are difficult to obtain with other microencapsulation techniques, aside from the high cell viability attained. Probiotic strains have been encapsulated onto protein or polysaccharide-based matrices by electrospraying, which involves the fragmentation of a liquid flow into droplets. In contrast to freeze-dried cells, microencapsulation via electrospraying in protein concentrate successfully extended the lifetime of bifidobacterial cells even under high relative humidity conditions. After cell cultivation, it has been discovered that subjecting cells to nutritional scarcity conditions can successfully induce cryo-tolerance in lactobacilli. By providing cross-protection against cold stresses, the adaptive signals discovered in response to starving conditions improved lactobacilli resilience to freezing. It's interesting to note that this cross-protection was accompanied by an increase in membrane moisture as well as a stress response that involves the upregulation of the proteins responsible for energy and carbohydrate metabolism as well as pH homeostasis. In particular, the mechanism of induced cross-protection entailed an increase in the level of fatty acid saturation and intracellular storage of reserve carbohydrates in the form of glycogen. It has been discovered that applying an acid adaption step before freeze-drying increases the cryo-tolerance of probiotic cells. The overexpression of proteins involved in energy metabolism and nucleotide synthesis, as well as the decline in the ratios of unsaturated to saturated and cyclic to saturated membrane fatty acids, were among the characteristics of the physiological cell responses in this situation.

When manufacturing and downstream processes are combined, the use of specific bioprocessing conditions must be taken into account. The fermentation conditions used during the cell manufacturing stage have a significant impact on cell development and resistance. Cell resistance to drying procedures can be greatly influenced by variables including pH, harvesting time, and fermentation temperature(Lacroix & Yildirim, 2007). The composition of the fermentation medium may contain defense mechanisms. By optimizing the bioprocessing conditions, it is possible to prevent the loss of acidification activity in some probiotic strains during storage at a chilled temperature. The cell resistance after processing and storage is significantly influenced by the harvesting time. Since the cells have a general stress tolerance, cells obtained during the stationary growth phase are often more resistant than cells harvested during the exponential development phase. Increased harvesting from the early to the late stationary growth phase in lactobacilli did not affect the culture quality in terms of biomass concentration. An advantage against loss of cell viability during dehydration operations can therefore be gained by optimizing the cell fermentation stage before cell concentration. Drying and rehydrating procedures in yeasts have been shown to cause metabolic activity loss in addition to membrane permeabilization. Harvesting yeast cells during their stationary growth phase has been found to mitigate these physiological alterations.

8.7 MONITORING

Monitoring and characterizing reactions at the single-cell level has also been the focus of research in the past ten years because both the cell survival ratio and physiological alterations play key roles in designing effective cell preservation procedures. Consequently, multiparameter flow cytometry has just recently emerged as a high-throughput tool to evaluate metabolic and structural alterations at the single-cell level, offering precise information on cell subpopulations that cannot be detected by conventional cell counting methods. Therefore, the use of this technique has made it possible to count the number of viable but uncultivable cells as well as to assess how freezing and frozen storage have affected the metabolic activity and membrane integrity of cells. Multiparameter flow cytometry has also been used to characterize factors like cell resistance of freeze-dried cells against acid stressors and the influence of adding various cryoprotectants to the freezing fluid. This high-throughput analytical method was also to assess the cryo-tolerance of LAB cultures both during and after freezing. Technical advancements like the use of larger size fluid nozzles to reduce mechanical stress during spray drying or the addition of sorbitol to prevent membrane-associated cell injuries in fluidized bed drying processes have been made possible by this improved understanding of the physiological responses of lactobacilli at the single-cell level.

Dehydration procedures like freeze-drying involve stressful conditions that have a negative effect on cellular viability. Furthermore, maintaining the complete integrity of the cellular membrane is only partially affected by tuning of operating variables like the freezing rate during freeze-drying operations. To lessen the negative effects caused by freezing processes, it may be possible to balance the ideal freezing rates and incorporate cryopreservatives into the freezing media. The effect of the osmotic response on cell viability during long-term preservation is another crucial factor. The long-term productive degeneration of microbial cell platforms, which can lead to subpar bio-production yields, is one of the fundamental problems of large-scale bio-based production. One of the causes of productivity failures during industrial bioprocesses is loss of stability and productive degeneration. Incorrect preservation techniques may result in undesirable phenotypic and physiological states in addition to the instability and loss of the productive metabolic capacity of the microbial cells during storage. The food sector depends on a continuous innovation process that results from the interaction of scientists, engineers, and consumers. Adsorption makes use of the capacity to stick to solid supports to create microbial biofilms, which can be as thin as a single layer or as thick as several millimeters. Strategies such as chemical crosslinking by glutaraldehyde, silica support, and metal oxide chelation can be used for microbial cells that do not adhere spontaneously.

Non-digestible carbohydrates, which are the basis for cell immobilization, are found in fruits. Cells that have been artificially immobilized, regardless of the method used, are not in their natural state. As a result, depending on the damage caused by the immobilization treatment, changes not only in cell viability but also in metabolic status may occur. A carbon source that is readily available to the filamentous fungus but not to the yeast should be present and supplied to the medium to encourage the development of yeast biocapsules. Thus, if the immobilization procedure is carried out properly, the yeast biocapsules are smooth, hollow spheres with walls made of fungus hyphae and yeast cells that may be reused in various fermentation processes without losing their integrity. The filamentous fungus was found to have been damaged during the fermentation process due to a lack of oxygen and only acted as an inert support, allowing the biocapsules to be reused in the future. In addition to increased tolerance to high substrate concentration, decreased end-product inhibition, major tolerance against toxic and inhibitory compounds, and stress tolerance, yeast biocapsules have also been shown to extend the activity and stability of the biocatalyst. These benefits make these biocapsules appealing for the manufacturing of ethanol and other alcoholic drinks like wine or beer.

Reference

1. Abd Elkader, A. M., Labib, S., Taha, T. F., Althobaiti, F., Aldhahrani, A., Salem, H. M., … Ibrahim, F. M. (2022). Phytogenic compounds from avocado (Persea americana L.) extracts; antioxidant activity, amylase inhibitory activity, therapeutic potential of type 2 diabetes. Saudi Journal of Biological Sciences, 29(3), 1428–1433. https://doi.org/10.1016/j.sjbs.2021.11.031
2. Akabanda, F., Owusu-Kwarteng, J., Tano-Debrah, K., Parkouda, C., & Jespersen, L. (2014). The Use of Lactic Acid Bacteria Starter Culture in the Production of Nunu , a Spontaneously Fermented Milk Product in Ghana. International Journal of Food Science, 2014, 1–11. https://doi.org/10.1155/2014/721067
3. Altermann, W., Kazmierczak, J., Oren, A., & Wright, D. T. (2006). Cyanobacterial calcification and its rock-building potential during 3.5 billion years of Earth history. Geobiology, 4(3), 147–166. https://doi.org/10.1111/j.1472-4669.2006.00076.x
4. Alvarez García, C. (2018). Chapter 11—Application of Enzymes for Fruit Juice Processing. In G. Rajauria & B. K. Tiwari (Eds.), Fruit Juices (pp. 201–216). San Diego: Academic Press. https://doi.org/10.1016/B978-0-12-802230-6.00011-4
5. Andersen, J. M., Barrangou, R., Abou Hachem, M., Lahtinen, S., Goh, Y. J., Svensson, B., & Klaenhammer, T. R. (2011). Transcriptional and functional analysis of galactooligosaccharide uptake by lacS in Lactobacillus acidophilus. Proceedings of the National Academy

of Sciences, 108(43), 17785–17790. https://doi.org/10.1073/pnas.1114152108

6. Archana, A., & Satyanarayana, T. (1997). Xylanase production by thermophilic Bacillus licheniformis A99 in solid-state fermentation. Enzyme and Microbial Technology, 21(1), 12–17. https://doi.org/10.1016/S0141-0229(96)00207-4
7. Ashaolu, T. J., & Reale, A. (2020). A Holistic Review on Euro-Asian Lactic Acid Bacteria Fermented Cereals and Vegetables. Microorganisms, 8(8), 1176. https://doi.org/10.3390/microorganisms8081176
8. Attri, S., & Goel, G. (2023). Chapter 22—Sterilization in bioprocesses. In A. K. Bhatt, R. K. Bhatia, & T. C. Bhalla (Eds.), Basic Biotechniques for Bioprocess and Bioentrepreneurship (pp. 329–339). Academic Press. https://doi.org/10.1016/B978-0-12-816109-8.00022-2
9. Bakratsas, G., Polydera, A., Katapodis, P., & Stamatis, H. (2021). Recent trends in submerged cultivation of mushrooms and their application as a source of nutraceuticals and food additives. Future Foods, 4, 100086. https://doi.org/10.1016/j.fufo.2021.100086
10. Bartowsky, E. J. (2005). Oenococcus oeni and malolactic fermentation – moving into the molecular arena. Australian Journal of Grape and Wine Research, 11(2), 174–187. https://doi.org/10.1111/j.1755-0238.2005.tb00286.x
11. Bartowsky, E. j., Costello, P. j., & Chambers, P. j. (2015). Emerging trends in the application of malolactic fermentation. Australian Journal of Grape and Wine Research, 21(S1), 663–669. https://doi.org/10.1111/ajgw.12185
12. Beasley, S. S., & Saris, P. E. J. (2004). Nisin-Producing Lactococcus lactis Strains Isolated from Human Milk. Applied and Environmental Microbiology, 70(8), 5051–5053. https://doi.org/10.1128/AEM.70.8.5051-5053.2004
13. Behera, S. S., Bal, P., Das, S., Panda, S. H., & Mohanty, N. (2018). Advances in Microbial Fermentation and Fermented Food for Health. In S. K. Panda & P. H. Shetty (Eds.), Innovations in Technologies for Fermented Food and Beverage Industries (pp. 53–69). Cham: Springer International Publishing. https://doi.org/10.1007/978-3-319-74820-7_3
14. Belton, B., Johnson, D. S., Thrift, E., Olsen, J., Hossain, M. A. R., & Thilsted, S. H. (2022). Dried fish at the intersection of food science,

economy, and culture: A global survey. Fish and Fisheries, 23(4), 941–962. https://doi.org/10.1111/faf.12664

15. Belz, M. C. E., Mairinger, R., Zannini, E., Ryan, L. A. M., Cashman, K. D., & Arendt, E. K. (2012). The effect of sourdough and calcium propionate on the microbial shelf-life of salt reduced bread. Applied Microbiology and Biotechnology, 96(2), 493–501. https://doi.org/10.1007/s00253-012-4052-x
16. Bevilacqua, A., Beneduce, L., Sinigaglia, M., & Corbo, M. R. (2013). Selection of Yeasts as Starter Cultures for Table Olives. Journal of Food Science, 78(5), M742–M751. https://doi.org/10.1111/1750-3841.12117
17. Bhanja Dey, T., & Banerjee, R. (2015). Purification, biochemical characterization and application of α-amylase produced by Aspergillus oryzae IFO-30103. Biocatalysis and Agricultural Biotechnology, 4(1), 83–90. https://doi.org/10.1016/j.bcab.2014.10.002
18. Bignell, D. R. D., Tahlan, K., Colvin, K. R., Jensen, S. E., & Leskiw, B. K. (2005). Expression of ccaR, Encoding the Positive Activator of Cephamycin C and Clavulanic Acid Production in Streptomyces clavuligerus, Is Dependent on bldG. Antimicrobial Agents and Chemotherapy, 49(4), 1529–1541. https://doi.org/10.1128/AAC.49.4.1529-1541.2005
19. Bonaccio, M., Iacoviello, L., de Gaetano, G., & on behalf of the Moli-sani Investigators. (2012). The Mediterranean diet: The reasons for a success. Thrombosis Research, 129(3), 401–404. https://doi.org/10.1016/j.thromres.2011.10.018
20. Borah, S., & Bhuyan, M. (2005). A computer based system for matching colours during the monitoring of tea fermentation. International Journal of Food Science & Technology, 40(6), 675–682. https://doi.org/10.1111/j.1365-2621.2005.00981.x
21. Breuer, U., & Harms, H. (2006). Debaryomyces hansenii—An extremophilic yeast with biotechnological potential. Yeast, 23(6), 415–437. https://doi.org/10.1002/yea.1374
22. Budak, H. N., & Guzel-Seydim, Z. B. (2010). Antioxidant activity and phenolic content of wine vinegars produced by two different techniques. Journal of the Science of Food and Agriculture, 90(12), 2021–2026. https://doi.org/10.1002/jsfa.4047

23. Buisson, G., Duée, E., Haser, R., & Payan, F. (1987). Three dimensional structure of porcine pancreatic alpha-amylase at 2.9 A resolution. Role of calcium in structure and activity. The EMBO Journal, 6(13), 3909–3916. https://doi.org/10.1002/j.1460-2075.1987.tb02731.x
24. Cahill, M. M. (1990). Bacterial flora of fishes: A review. Microbial Ecology, 19(1), 21–41. https://doi.org/10.1007/BF02015051
25. Campbell-Platt, G. (1994). Fermented foods—A world perspective. Food Research International, 27(3), 253–257. https://doi.org/10.1016/0963-9969(94)90093-0
26. Canete-Rodriguez, A. M., Santos-Duenas, I. M., Jimenez-Hornero, J. E., Ehrenreich, A., Liebl, W., & García-García, I. (2016). Gluconic acid: Properties, production methods and applications—An excellent opportunity for agro-industrial by-products and waste bio-valorization. Process Biochemistry, 51(12), 1891–1903. https://doi.org/10.1016/j.procbio.2016.08.028
27. Casertano, M., Fogliano, V., & Ercolini, D. (2022). Psychobiotics, gut microbiota and fermented foods can help preserving mental health. Food Research International, 152, 110892. https://doi.org/10.1016/j.foodres.2021.110892
28. Cavalieri, D., McGovern, P. E., Hartl, D. L., Mortimer, R., & Polsinelli, M. (2003). Evidence for S. cerevisiae Fermentation in Ancient Wine. Journal of Molecular Evolution, 57(1), S226–S232. https://doi.org/10.1007/s00239-003-0031-2
29. Chae, P. S., Rasmussen, S. G. F., Rana, R. R., Gotfryd, K., Chandra, R., Goren, M. A., … Gellman, S. H. (2010). Maltose–neopentyl glycol (MNG) amphiphiles for solubilization, stabilization and crystallization of membrane proteins. Nature Methods, 7(12), 1003–1008. https://doi.org/10.1038/nmeth.1526
30. Chao, S.-H., Wu, R.-J., Watanabe, K., & Tsai, Y.-C. (2009). Diversity of lactic acid bacteria in suan-tsai and fu-tsai, traditional fermented mustard products of Taiwan. International Journal of Food Microbiology, 135(3), 203–210. https://doi.org/10.1016/j.ijfoodmicro.2009.07.032
31. Chatterjee, M., Chakrabarty, S. L., Chattopadhyay, B. D., & Mandal, R. K. (1997). Production of lactic acid by direct fermentation of starchy wastes by an amylase-producing Lactobacillus. Biotechnology Letters, 19(9), 873–874. https://doi.org/10.1023/A:1018337604225
32. Cheigh, H., Park, K., & Lee, C. Y. (1994). Biochemical, microbiological, and nutritional aspects of kimchi (Korean fermented vegetable

products). Critical Reviews in Food Science and Nutrition, 34(2), 175–203. https://doi.org/10.1080/10408399409527656

33. Christensen, L. F., García-Béjar, B., Bang-Berthelsen, C. H., & Hansen, E. B. (2022). Extracellular microbial proteases with specificity for plant proteins in food fermentation. International Journal of Food Microbiology, 381, 109889. https://doi.org/10.1016/j.ijfoodmicro.2022.109889
34. Christopher, L. P., Hemanathan Kumar, & Zambare, V. P. (2014). Enzymatic biodiesel: Challenges and opportunities. Applied Energy, 119, 497–520. https://doi.org/10.1016/j.apenergy.2014.01.017
35. Chun, L. P., Miguel, M. J., Junkins, E. N., Forbes, S. L., & Carter, D. O. (2015). An initial investigation into the ecology of culturable aerobic postmortem bacteria. Science & Justice, 55(6), 394–401. https://doi.org/10.1016/j.scijus.2015.07.003
36. Coulter, S. T., & Combs, W. B. (1939). The Use of Steapsin in the Manufacture of Blue Cheese*. Journal of Dairy Science, 22(7), 521–525. https://doi.org/10.3168/jds.S0022-0302(39)92908-5
37. Couteau, D., McCartney, A. L., Gibson, G. R., Williamson, G., & Faulds, C. B. (2001). Isolation and characterization of human colonic bacteria able to hydrolyse chlorogenic acid. Journal of Applied Microbiology, 90(6), 873–881. https://doi.org/10.1046/j.1365-2672.2001.01316.x
38. Curic, M., de Richelieu, M., Henriksen, C. M., Jochumsen, K. V., Villadsen, J., & Nilsson, D. (1999). Glucose/Citrate Cometabolism in Lactococcus lactis subsp. Lactis biovar diacetylactis with Impaired α-Acetolactate Decarboxylase. Metabolic Engineering, 1(4), 291–298. https://doi.org/10.1006/mben.1999.0123
39. De Buyser, D. R., Spriet, J. A., & Vansteenkiste, G. C. (1985). Computer-Control of Fermentation Processes. In N. G. Alaeddinoğlu, A. L. Demain, & G. Lancini (Eds.), Industrial Aspects of Biochemistry and Genetics (pp. 189–236). Boston, MA: Springer US. https://doi.org/10.1007/978-1-4684-1227-7_13
40. Dedey, K. B., Grenier, D., & Lucas, T. (2021). Mathematical modelling of uniaxial extension of a heterogeneous gas cell wall in bread dough: Stress fields and stress concentration analysis relating to the proving and baking steps. Journal of Food Engineering, 308, 110669. https://doi.org/10.1016/j.jfoodeng.2021.110669

41. Demirbas, A. (2007). Producing and Using Bioethanol as an Automotive Fuel. Energy Sources, Part B: Economics, Planning, and Policy, 2(4), 391–401. https://doi.org/10.1080/15567240600705466
42. Demirkaya, E., Ciftcioglu, B., Ozyildiz, G., Zengin, G. E., Pala-Ozkok, I., Cokgor, E., & Tas, D. O. (2022). Comprehensive evaluation of starter culture impact on the bioreactor performance and microbial kinetics. Biochemical Engineering Journal, 177, 108233. https://doi.org/10.1016/j.bej.2021.108233
43. Deng, Y., Misselwitz, B., Dai, N., & Fox, M. (2015). Lactose Intolerance in Adults: Biological Mechanism and Dietary Management. Nutrients, 7(9), 8020–8035. https://doi.org/10.3390/nu7095380
44. Dennis, C., & Blijham, J. M. (1980). Effect of temperature on viability of sporangiospores of Rhizopus and Mucor species. Transactions of the British Mycological Society, 74(1), 89–94. https://doi.org/10.1016/S0007-1536(80)80012-X
45. Desai, K. M., Survase, S. A., Saudagar, P. S., Lele, S. S., & Singhal, R. S. (2008). Comparison of artificial neural network (ANN) and response surface methodology (RSM) in fermentation media optimization: Case study of fermentative production of scleroglucan. Biochemical Engineering Journal, 41(3), 266–273. https://doi.org/10.1016/j.bej.2008.05.009
46. Dhillon, G. S., Oberoi, H. S., Kaur, S., Bansal, S., & Brar, S. K. (2011). Value-addition of agricultural wastes for augmented cellulase and xylanase production through solid-state tray fermentation employing mixed-culture of fungi. Industrial Crops and Products, 34(1), 1160–1167. https://doi.org/10.1016/j.indcrop.2011.04.001
47. DICKS, L. M. T., DELLAGLIO, F., & COLLINS, M. D. (1995). Proposal To Reclassify Leuconostoc oenos as Oenococcus oeni [corrig.] gen. Nov., comb. Nov. International Journal of Systematic and Evolutionary Microbiology, 45(2), 395–397. https://doi.org/10.1099/00207713-45-2-395
48. Ekpong, B. (2009). Effects of seed maturity, seed storage and pre-germination treatments on seed germination of cleome (Cleome gynandra L.). Scientia Horticulturae, 119(3), 236–240. https://doi.org/10.1016/j.scienta.2008.08.003
49. Etemadian, Y., Shabanpour, B., Sadeghi Mahoonak, A. R., Shabani, A., & Alami, M. (2011). Cryoprotective effects of polyphosphates on Rutilus frisii kutum fillets during ice storage. Food Chemistry, 129(4), 1544–1551. https://doi.org/10.1016/j.foodchem.2011.06.005

50. Faris, J. D. (2014). Wheat Domestication: Key to Agricultural Revolutions Past and Future. In R. Tuberosa, A. Graner, & E. Frison (Eds.), Genomics of Plant Genetic Resources: Volume 1. Managing, sequencing and mining genetic resources (pp. 439–464). Dordrecht: Springer Netherlands. https://doi.org/10.1007/978-94-007-7572-5_18
51. Farooq, M. A., Ali, S., Hassan, A., Tahir, H. M., Mumtaz, S., & Mumtaz, S. (2021). Biosynthesis and industrial applications of α-amylase: A review. Archives of Microbiology, 203(4), 1281–1292. https://doi.org/10.1007/s00203-020-02128-y
52. Feutry, F., Torre, P., Arana, I., Garcia, S., Pérez Elortondo, F. J., & Berthier, F. (2016). Suitability of a new mixed-strain starter for manufacturing uncooked raw ewe's milk cheeses. Food Microbiology, 56, 52–68. https://doi.org/10.1016/j.fm.2015.12.004
53. Formenti, L. R., Nørregaard, A., Bolic, A., Hernandez, D. Q., Hagemann, T., Heins, A.-L., … Gernaey, K. V. (2014). Challenges in industrial fermentation technology research. Biotechnology Journal, 9(6), 727–738. https://doi.org/10.1002/biot.201300236
54. Gallazzi, A., Branska, B., Marinelli, F., & Patakova, P. (2015). Continuous production of n-butanol by Clostridium pasteurianum DSM 525 using suspended and surface-immobilized cells. Journal of Biotechnology, 216, 29–35. https://doi.org/10.1016/j.jbiotec.2015.10.008
55. Gao, Y., Hou, L., Gao, J., Li, D., Tian, Z., Fan, B., … Li, S. (2021). Metabolomics Approaches for the Comprehensive Evaluation of Fermented Foods: A Review. Foods, 10(10), 2294. https://doi.org/10.3390/foods10102294
56. Garcia-Ochoa, F., Gomez, E., Santos, V. E., & Merchuk, J. C. (2010). Oxygen uptake rate in microbial processes: An overview. Biochemical Engineering Journal, 49(3), 289–307. https://doi.org/10.1016/j.bej.2010.01.011
57. Garrity, G. M., & Holt, J. G. (2001). The Road Map to the Manual. In D. R. Boone, R. W. Castenholz, & G. M. Garrity (Eds.), Bergey's Manual® of Systematic Bacteriology: Volume One: The Archaea and the Deeply Branching and Phototrophic Bacteria (pp. 119–166). New York, NY: Springer. https://doi.org/10.1007/978-0-387-21609-6_15
58. Gould, G. W. (1996). Methods for preservation and extension of shelf life. International Journal of Food Microbiology, 33(1), 51–64. https://doi.org/10.1016/0168-1605(96)01133-6

59. Guzmán-Maldonado, H., Paredes-López, O., & Biliaderis, C. G. (1995). Amylolytic enzymes and products derived from starch: A review. Critical Reviews in Food Science and Nutrition, 35(5), 373–403. https://doi.org/10.1080/10408399509527706
60. Hachmeister, K. A., & Fung, D. Y. C. (1993). Tempeh: A Mold-Modified Indigenous Fermented Food Made from Soybeans and/or Cereal Grains. Critical Reviews in Microbiology, 19(3), 137–188. https://doi.org/10.3109/10408419309113527
61. Handoyo, T., & Morita, N. (2006). Structural and Functional Properties of Fermented Soybean (Tempeh) by Using Rhizopus oligosporus. International Journal of Food Properties, 9(2), 347–355. https://doi.org/10.1080/10942910500224746
62. Hansen, G. H., Lübeck, M., Frisvad, J. C., Lübeck, P. S., & Andersen, B. (2015). Production of cellulolytic enzymes from ascomycetes: Comparison of solid state and submerged fermentation. Process Biochemistry, 50(9), 1327–1341. https://doi.org/10.1016/j.procbio.2015.05.017
63. Haque, Md. A., Lee, J. H., & Cho, K. M. (2015). Endophytic bacterial diversity in Korean kimchi made of Chinese cabbage leaves and their antimicrobial activity against pathogens. Food Control, 56, 24–33. https://doi.org/10.1016/j.foodcont.2015.03.006
64. Hartmanis, M. G. N., & Gatenbeck, S. (1984). Intermediary Metabolism in Clostridium acetobutylicum: Levels of Enzymes Involved in the Formation of Acetate and Butyrate. Applied and Environmental Microbiology, 47(6), 1277–1283. https://doi.org/10.1128/aem.47.6.1277-1283.1984
65. He, Z. H., Liu, A. H., Javier Pena, R., & Rajaram, S. (2003). Suitability of Chinese wheat cultivars for production of northern style Chinese steamed bread. Euphytica, 131(2), 155–163. https://doi.org/10.1023/A:1023929513167
66. Hernandez-Orte, P., Lapena, A. C., Pena-Gallego, A., Astrain, J., Baron, C., Pardo, I. & Ferreira, V. (2008). Biogenic amine determination in wine fermented in oak barrels: Factors affecting formation. Food Research International, 41(7), 697–706. https://doi.org/10.1016/j.foodres.2008.05.002
67. Holker, U., Hofer, M., & Lenz, J. (2004). Biotechnological advantages of laboratory-scale solid-state fermentation with fungi. Applied Microbiology and Biotechnology, 64(2), 175–186. https://doi.org/10.1007/s00253-003-1504-3

68. Holzapfel, W. H. (2002). Appropriate starter culture technologies for small-scale fermentation in developing countries. International Journal of Food Microbiology, 75(3), 197–212. https://doi.org/10.1016/S0168-1605(01)00707-3

69. Houbraken, J., Frisvad, J. C., & Samson, R. A. (2011). Fleming's penicillin producing strain is not Penicillium chrysogenum but P. rubens. IMA Fungus, 2(1), 87–95. https://doi.org/10.5598/imafungus.2011.02.01.12

70. Hui, Y. H., & Evranuz, E. Ö. (Eds.). (2012). Handbook of Fermented Food and Beverage Technology Two Volume Set (2nd ed.). Boca Raton: CRC Press. https://doi.org/10.1201/b15438

71. Imandi, S. B., Bandaru, V. V. R., Somalanka, S. R., Bandaru, S. R., & Garapati, H. R. (2008). Application of statistical experimental designs for the optimization of medium constituents for the production of citric acid from pineapple waste. Bioresource Technology, 99(10), 4445–4450. https://doi.org/10.1016/j.biortech.2007.08.071

72. Ito, J., Nakagawa, K., Kato, S., Hirokawa, T., Kuwahara, S., Nagai, T., & Miyazawa, T. (2016). A novel chiral stationary phase HPLC-MS/MS method to discriminate between enzymatic oxidation and auto-oxidation of phosphatidylcholine. Analytical and Bioanalytical Chemistry, 408(27), 7785–7793. https://doi.org/10.1007/s00216-016-9882-4

73. Jarocki, P., Komoń-Janczara, E., Glibowska, A., Dworniczak, M., Pytka, M., Korzeniowska-Kowal, A., … Kordowska-Wiater, M. (2020). Molecular Routes to Specific Identification of the Lactobacillus Casei Group at the Species, Subspecies and Strain Level. International Journal of Molecular Sciences, 21(8), 2694. https://doi.org/10.3390/ijms21082694

74. Jeyaram, K., Mohendro Singh, W., Premarani, T., Devi, A. R., Chanu, K. S., Talukdar, N. C., & Singh, M. R. (2008). Molecular identification of dominant microflora associated with 'Hawaijar'—A traditional fermented soybean (Glycine max (L.)) food of Manipur, India. International Journal of Food Microbiology, 122(3), 259–268. https://doi.org/10.1016/j.ijfoodmicro.2007.12.026

75. Jurus, A. M., & Sundberg, W. J. (1976). Penetration of Rhizopus oligosporus into Soybeans in Tempeh. Applied and Environmental Microbiology, 32(2), 284–287. https://doi.org/10.1128/aem.32.2.284-287.1976

76. Juvonen, R., Honkapää, K., Maina, N. H., Shi, Q., Viljanen, K., Maaheimo, H., … Lantto, R. (2015). The impact of fermentation with exopolysaccharide producing lactic acid bacteria on rheological, chemical and sensory properties of pureed carrots (Daucus carota L.). International Journal of Food Microbiology, 207, 109–118. https://doi.org/10.1016/j.ijfoodmicro.2015.04.031
77. Kango, N., Jana, U. K., & Choukade, R. (2019). Fungal Enzymes: Sources and Biotechnological Applications. In T. Satyanarayana, S. K. Deshmukh, & M. V. Deshpande (Eds.), Advancing Frontiers in Mycology & Mycotechnology: Basic and Applied Aspects of Fungi (pp. 515–538). Singapore: Springer. https://doi.org/10.1007/978-981-13-9349-5_21
78. Kennedy, M., & Krouse, D. (1999). Strategies for improving fermentation medium performance: A review. Journal of Industrial Microbiology and Biotechnology, 23(6), 456–475. https://doi.org/10.1038/sj.jim.2900755
79. Kerwin, J. L. (2007). Oomycetes: Lagendium Giganteum. Journal of the American Mosquito Control Association, 23(sp2), 50–57. https://doi.org/10.2987/8756-971X(2007)23[50:OLG]2.0.CO;2
80. Khajeh, K., Shokri, M. M., Asghari, S. M., Moradian, F., Ghasemi, A., Sadeghi, M., … Naderi-Manesh, H. (2006). Acidic and proteolytic digestion of α-amylases from Bacillus licheniformis and Bacillus amyloliquefaciens: Stability and flexibility analysis. Enzyme and Microbial Technology, 38(3), 422–428. https://doi.org/10.1016/j.enzmictec.2005.06.021
81. Khemakhem, B., Ali, M. B., Aghajari, N., Juy, M., Haser, R., & Bejar, S. (2009). Engineering of the α-amylase from Geobacillus stearothermophilus US100 for detergent incorporation. Biotechnology and Bioengineering, 102(2), 380–389. https://doi.org/10.1002/bit.22083
82. Kim, D.-H., Jang, S., Yun, Y.-M., Lee, M.-K., Moon, C., Kang, W.-S., … Kim, M.-S. (2014). Effect of acid-pretreatment on hydrogen fermentation of food waste: Microbial community analysis by next generation sequencing. International Journal of Hydrogen Energy, 39(29), 16302–16309. https://doi.org/10.1016/j.ijhydene.2014.08.004
83. Kim, M., & Chun, J. (2005). Bacterial community structure in kimchi, a Korean fermented vegetable food, as revealed by 16S rRNA gene analysis. International Journal of Food Microbiology, 103(1), 91–96. https://doi.org/10.1016/j.ijfoodmicro.2004.11.030

84. Kleerebezemab, M., Hols, P., & Hugenholtz, J. (2000). Lactic acid bacteria as a cell factory: Rerouting of carbon metabolism in Lactococcus lactis by metabolic engineering. Enzyme and Microbial Technology, 26(9), 840–848. https://doi.org/10.1016/S0141-0229(00)00180-0

85. Kong, Q., Chi, C., Yu, J., Shan, S., Li, Q., Li, Q., … Bennett, J. W. (2014). The inhibitory effect of Bacillus megaterium on aflatoxin and cyclopiazonic acid biosynthetic pathway gene expression in Aspergillus flavus. Applied Microbiology and Biotechnology, 98(11), 5161–5172. https://doi.org/10.1007/s00253-014-5632-8

86. Kovacs, K. L., Bagyinka, C., & Tigyi, G. (1988). Proteolytic resistance and its utilization in purification of hydrogenase from Thiocapsa roseopersicina. Biochimica et Biophysica Acta (BBA) - Bioenergetics, 935(2), 166–172. https://doi.org/10.1016/0005-2728(88)90213-7

87. Kuleasan, H., & Çakmakci, M. L. (2002). Effect of reuterin, produced by Lactobacillus reuteri on the surface of sausages to inhibit the growth of Listeria monocytogenes and Salmonella spp. Food / Nahrung, 46(6), 408–410. https://doi.org/10.1002/1521-3803(20021101)46:6<408::AID-FOOD408>3.0.CO;2-T

88. Kumar, P. K. R., & Lonsane, B. K. (1987). Gibberellic acid by solid state fermentation: Consistent and improved yields. Biotechnology and Bioengineering, 30(2), 267–271. https://doi.org/10.1002/bit.260300217

89. Laan, H., Smid, E. J., de Leij, L., Schwander, E., & Konings, W. N. (1988). Monoclonal Antibodies to the Cell-Wall-Associated Proteinase of Lactococcus lactis subsp. Cremoris Wg2. Applied and Environmental Microbiology, 54(9), 2250–2256. https://doi.org/10.1128/aem.54.9.2250-2256.1988

90. Lacroix, C., & Yildirim, S. (2007). Fermentation technologies for the production of probiotics with high viability and functionality. Current Opinion in Biotechnology, 18(2), 176–183. https://doi.org/10.1016/j.copbio.2007.02.002

91. Landaud, S., Helinck, S., & Bonnarme, P. (2008). Formation of volatile sulfur compounds and metabolism of methionine and other sulfur compounds in fermented food. Applied Microbiology and Biotechnology, 77(6), 1191–1205. https://doi.org/10.1007/s00253-007-1288-y

92. Legras, J.-L., Merdinoglu, D., Cornuet, J.-M., & Karst, F. (2007). Bread, beer and wine: Saccharomyces cerevisiae diversity reflects

human history. Molecular Ecology, 16(10), 2091–2102. https://doi.org/10.1111/j.1365-294X.2007.03266.x

93. Liese, A., & Hilterhaus, L. (2013). Evaluation of immobilized enzymes for industrial applications. Chemical Society Reviews, 42(15), 6236–6249. https://doi.org/10.1039/C3CS35511J
94. Linko, Y.-Y., Javanainen, P., & Linko, S. (1997). Biotechnology of bread baking. Trends in Food Science & Technology, 8(10), 339–344. https://doi.org/10.1016/S0924-2244(97)01066-2
95. Liu, S.-Q., & Pilone, G. J. (2000). An overview of formation and roles of acetaldehyde in winemaking with emphasis on microbiological implications. International Journal of Food Science & Technology, 35(1), 49–61. https://doi.org/10.1046/j.1365-2621.2000.00341.x
96. Liu, Y.C., Wang, F.S., & Lee, W.C. (2001). On-line monitoring and controlling system for fermentation processes. Biochemical Engineering Journal, 7(1), 17–25. https://doi.org/10.1016/S1369-703X(00)00100-5
97. Lopez-Casado, G., Urbanowicz, B. R., Damasceno, C. M., & Rose, J. K. (2008). Plant glycosyl hydrolases and biofuels: A natural marriage. Current Opinion in Plant Biology, 11(3), 329–337. https://doi.org/10.1016/j.pbi.2008.02.010
98. Lowe, D. P., & Arendt, E. K. (2004). The Use and Effects of Lactic Acid Bacteria in Malting and Brewing with Their Relationships to Antifungal Activity, Mycotoxins and Gushing: A Review. Journal of the Institute of Brewing, 110(3), 163–180. https://doi.org/10.1002/j.2050-0416.2004.tb00199.x
99. Lubbers, S., Decourcelle, N., Vallet, N., & Guichard, E. (2004). Flavor Release and Rheology Behavior of Strawberry Fatfree Stirred Yogurt during Storage. Journal of Agricultural and Food Chemistry, 52(10), 3077–3082. https://doi.org/10.1021/jf0352374
100. MacLeod, M., & Nersessian, N. J. (2015). Modeling systems-level dynamics: Understanding without mechanistic explanation in integrative systems biology. Studies in History and Philosophy of Science Part C: Studies in History and Philosophy of Biological and Biomedical Sciences, 49, 1–11. https://doi.org/10.1016/j.shpsc.2014.10.004
101. Maeda, T., Hashimoto, T., Minoda, M., Tamagawa, S., & Morita, N. (2003). Effects of Mutant Thermostable α-Amylases on Rheological Properties of Wheat Dough and Bread. Cereal Chemistry, 80(6), 722–

727. https://doi.org/10.1094/CCHEM.2003.80.6.722

102. Magyar, I., & Tóth, T. (2011). Comparative evaluation of some oenological properties in wine strains of Candida stellata, Candida zemplinina, Saccharomyces uvarum and Saccharomyces cerevisiae. Food Microbiology, 28(1), 94–100. https://doi.org/10.1016/j.fm.2010.08.011
103. Malosso, E., Waite, I. S., English, L., Hopkins, D. W., & O'Donnell, A. G. (2006). Fungal diversity in maritime Antarctic soils determined using a combination of culture isolation, molecular fingerprinting and cloning techniques. Polar Biology, 29(7), 552–561. https://doi.org/10.1007/s00300-005-0088-z
104. Marco, M. L., Sanders, M. E., Gänzle, M., Arrieta, M. C., Cotter, P. D., De Vuyst, L., … Hutkins, R. (2021). The International Scientific Association for Probiotics and Prebiotics (ISAPP) consensus statement on fermented foods. Nature Reviews Gastroenterology & Hepatology, 18(3), 196–208. https://doi.org/10.1038/s41575-020-00390-5
105. Martinez, M. M., Li, C., Okoniewska, M., Mukherjee, I., Vellucci, D., & Hamaker, B. (2018). Slowly digestible starch in fully gelatinized material is structurally driven by molecular size and A and B1 chain lengths. Carbohydrate Polymers, 197, 531–539. https://doi.org/10.1016/j.carbpol.2018.06.021
106. Masih, E. I., & Paul, B. (2002). Secretion of β-1,3-Glucanases by the Yeast Pichia membranifaciens and Its Possible Role in the Biocontrol of Botrytis cinerea Causing Grey Mold Disease of the Grapevine. Current Microbiology, 44(6), 391–395. https://doi.org/10.1007/s00284-001-0011-y
107. Menezes, A. G. T., Ramos, C. L., Dias, D. R., & Schwan, R. F. (2018). Combination of probiotic yeast and lactic acid bacteria as starter culture to produce maize-based beverages. Food Research International, 111, 187–197. https://doi.org/10.1016/j.foodres.2018.04.065
108. Mirás-Avalos, J. M., & Araujo, E. S. (2021). Optimization of Vineyard Water Management: Challenges, Strategies, and Perspectives. Water, 13(6), 746. https://doi.org/10.3390/w13060746
109. Mohammadi, M., Younesi, H., Najafpour, G., & Mohamed, A. R. (2012). Sustainable ethanol fermentation from synthesis gas by Clostridium ljungdahlii in a continuous stirred tank bioreactor. Journal of Chemical Technology & Biotechnology, 87(6), 837–843. https://doi.org/10.1002/jctb.3712

110. Mullan, W. M. A. (2000). Causes and control of early gas production in cheddar cheese. International Journal of Dairy Technology, 53(2), 63–68. https://doi.org/10.1111/j.1471-0307.2000.tb02660.x
111. Nout, M. J. R. (1994). Fermented foods and food safety. Food Research International, 27(3), 291–298. https://doi.org/10.1016/0963-9969(94)90097-3
112. Obafemi, Y. D., Oranusi, S. U., Ajanaku, K. O., Akinduti, P. A., Leech, J., & Cotter, P. D. (2022). African fermented foods: Overview, emerging benefits, and novel approaches to microbiome profiling. Npj Science of Food, 6(1), 15. https://doi.org/10.1038/s41538-022-00130-w
113. Oerke, E.-C., & Dehne, H.-W. (1997). Global crop production and the efficacy of crop protection—Current situation and future trends. European Journal of Plant Pathology, 103(3), 203–215. https://doi.org/10.1023/A:1008602111248
114. Oh, S.-E., Zuo, Y., Zhang, H., Guiltinan, M. J., Logan, B. E., & Regan, J. M. (2009). Hydrogen production by Clostridium acetobutylicum ATCC 824 and megaplasmid-deficient mutant M5 evaluated using a large headspace volume technique. International Journal of Hydrogen Energy, 34(23), 9347–9353. https://doi.org/10.1016/j.ijhydene.2009.09.084
115. Oliveira, P. M., Zannini, E., & Arendt, E. K. (2014). Cereal fungal infection, mycotoxins, and lactic acid bacteria mediated bioprotection: From crop farming to cereal products. Food Microbiology, 37, 78–95. https://doi.org/10.1016/j.fm.2013.06.003
116. Olympia, M., Fukuda, H., Ono, H., Kaneko, Y., & Takano, M. (1995). Characterization of starch-hydrolyzing lactic acid bacteria isolated from a fermented fish and rice food, "burong isda", and its amylolytic enzyme. Journal of Fermentation and Bioengineering, 80(2), 124–130. https://doi.org/10.1016/0922-338X(95)93206-Y
117. O'Sullivan, E., & Condon, S. (1997). Intracellular pH is a major factor in the induction of tolerance to acid and other stresses in Lactococcus lactis. Applied and Environmental Microbiology, 63(11), 4210–4215. https://doi.org/10.1128/aem.63.11.4210-4215.1997
118. Pan, H., & Sun, M.-Y. (2011). Variations of alkenone based paleotemperature index (U37K′) during Emiliania huxleyi cell growth, respiration (auto-metabolism) and microbial degradation. Organic Geochemistry, 42(6), 678–687. https://doi.org/10.1016/j.orggeochem.2011.03.024

119. Park, K.-Y., Jung, K.-O., Rhee, S.-H., & Choi, Y. H. (2003). Antimutagenic effects of doenjang (Korean fermented soypaste) and its active compounds. Mutation Research/Fundamental and Molecular Mechanisms of Mutagenesis, 523–524, 43–53. https://doi.org/10.1016/S0027-5107(02)00320-2

120. Paulose, S. K., & Chakraborty, K. (2021). Antioxidant spiropharanone, an undescribed variant of trans-decalin spiro-γ-lactone, from pharaoh cuttlefish Sepia pharaonis: Twin inhibitors of inflammatory 5-lipoxygenase and serine protease dipeptidyl peptidase-4. Journal of Food Biochemistry, 45(10), e13919. https://doi.org/10.1111/jfbc.13919

121. Peñas, E., Martinez-Villaluenga, C., & Frias, J. (2017). Chapter 24 - Sauerkraut: Production, Composition, and Health Benefits. In J. Frias, C. Martinez-Villaluenga, & E. Peñas (Eds.), Fermented Foods in Health and Disease Prevention (pp. 557–576). Boston: Academic Press. https://doi.org/10.1016/B978-0-12-802309-9.00024-8

122. Pinto, M. F., Ponsano, E. H. G., Franco, B. D. G. M., & Shimokomaki, M. (2002). Charqui meats as fermented meat products: Role of bacteria for some sensorial properties development. Meat Science, 61(2), 187–191. https://doi.org/10.1016/S0309-1740(01)00184-X

123. Prakash, O., Nimonkar, Y., & Desai, D. (2020). A Recent Overview of Microbes and Microbiome Preservation. Indian Journal of Microbiology, 60(3), 297–309. https://doi.org/10.1007/s12088-020-00880-9

124. Pretorius, I. S. (2000). Tailoring wine yeast for the new millennium: Novel approaches to the ancient art of winemaking. Yeast, 16(8), 675–729. https://doi.org/10.1002/1097-0061(20000615)16:8<675::AID-YEA585>3.0.CO;2-B

125. Pua, A., Tang, V. C. Y., Goh, R. M. V., Sun, J., Lassabliere, B., & Liu, S. Q. (2022). Ingredients, Processing, and Fermentation: Addressing the Organoleptic Boundaries of Plant-Based Dairy Analogues. Foods, 11(6), 875. https://doi.org/10.3390/foods11060875

126. Pulz, O. (2001). Photobioreactors: Production systems for phototrophic microorganisms. Applied Microbiology and Biotechnology, 57(3), 287–293. https://doi.org/10.1007/s002530100702

127. Quiros, A., Hernandez-Ledesma, B., Ramos, M., Amigo, L., & Recio, I. (2005). Angiotensin-Converting Enzyme Inhibitory Activity of Peptides Derived from Caprine Kefir. Journal of Dairy Science, 88(10), 3480–3487. https://doi.org/10.3168/jds.S0022-0302(05)73032-0

128. Rainieri, S., Kodama, Y., Kaneko, Y., Mikata, K., Nakao, Y., & Ashikari, T. (2006). Pure and Mixed Genetic Lines of Saccharomyces bayanus and Saccharomyces pastorianus and Their Contribution to the Lager Brewing Strain Genome. Applied and Environmental Microbiology, 72(6), 3968–3974. https://doi.org/10.1128/AEM.02769-05
129. Raman, J., Kim, J.-S., Choi, K. R., Eun, H., Yang, D., Ko, Y.-J., & Kim, S.-J. (2022). Application of Lactic Acid Bacteria (LAB) in Sustainable Agriculture: Advantages and Limitations. International Journal of Molecular Sciences, 23(14), 7784. https://doi.org/10.3390/ijms23147784
130. Rao, R., & Basak, N. (2021). Process optimization and mathematical modelling of photo-fermentative hydrogen production from dark fermentative cheese whey effluent by Rhodobacter sphaeroides O.U.001 in 2-L cylindrical bioreactor. Biomass Conversion and Biorefinery. https://doi.org/10.1007/s13399-021-01377-1
131. Reddy, G., Altaf, Md., Naveena, B. J., Venkateshwar, M., & Kumar, E. V. (2008). Amylolytic bacterial lactic acid fermentation—A review. Biotechnology Advances, 26(1), 22–34. https://doi.org/10.1016/j.biotechadv.2007.07.004
132. Richards, T. A., Leonard, G., & Wideman, J. G. (2017). What Defines the "Kingdom" Fungi? Microbiology Spectrum, 5(3), 5.3.23. https://doi.org/10.1128/microbiolspec.FUNK-0044-2017
133. Robinson, T., Singh, D., & Nigam, P. (2001). Solid-state fermentation: A promising microbial technology for secondary metabolite production. Applied Microbiology and Biotechnology, 55(3), 284–289. https://doi.org/10.1007/s002530000565
134. Ropars, J., Didiot, E., Rodríguez de la Vega, R. C., Bennetot, B., Coton, M., Poirier, E., … Giraud, T. (2020). Domestication of the Emblematic White Cheese-Making Fungus Penicillium camemberti and Its Diversification into Two Varieties. Current Biology, 30(22), 4441-4453.e4. https://doi.org/10.1016/j.cub.2020.08.082
135. Saini, R., Saini, H. S., & Dahiya, A. (2017). Amylases: Characteristics and industrial applications. Journal of Pharmacognosy and Phytochemistry, 6(4), 1865–1871. Retrieved from https://www.phytojournal.com/archives/2017.v6.i4.1586/amylases-characteristics-and-industrial-applications
136. Saqib, S., Akram, A., Halim, S. A., & Tassaduq, R. (2017). Sources of β-galactosidase and its applications in food industry. 3 Biotech, 7(1), 79. https://doi.org/10.1007/s13205-017-0645-5

137. Sarangi, P. K., & Nanda, S. (2018). Recent Developments and Challenges of Acetone-Butanol-Ethanol Fermentation. In Prakash Kumar Sarangi, S. Nanda, & P. Mohanty (Eds.), Recent Advancements in Biofuels and Bioenergy Utilization (pp. 111–123). Singapore: Springer. https://doi.org/10.1007/978-981-13-1307-3_5

138. Sarkar, T., Salauddin, M., Roy, A., Sharma, N., Sharma, A., Yadav, S., … Simal-Gandara, J. (2022). Minor tropical fruits as a potential source of bioactive and functional foods. Critical Reviews in Food Science and Nutrition, 0(0), 1–45. https://doi.org/10.1080/10408398.2022.2033953

139. Sawadogo-Lingani, H., Owusu-Kwarteng, J., Glover, R., Diawara, B., Jakobsen, M., & Jespersen, L. (2021). Sustainable Production of African Traditional Beers With Focus on Dolo, a West African Sorghum-Based Alcoholic Beverage. Frontiers in Sustainable Food Systems, 5. Retrieved from https://www.frontiersin.org/articles/10.3389/fsufs.2021.672410

140. Saxena, R. K., Gupta, R., Saxena, S., & Gulati, R. (2001). Role of fungal enzymes in food processing. In G. G. Khachatourians & D. K. Arora (Eds.), Applied Mycology and Biotechnology (pp. 353–386). Elsevier. https://doi.org/10.1016/S1874-5334(01)80015-0

141. Schindelin, H., Marahiel, M. A., & Heinemann, U. (1993). Universal nucleic acid-binding domain revealed by crystal structure of the B. subtilis major cold-shock protein. Nature, 364(6433), 164–168. https://doi.org/10.1038/364164a0

142. Selhub, E. M., Logan, A. C., & Bested, A. C. (2014). Fermented foods, microbiota, and mental health: Ancient practice meets nutritional psychiatry. Journal of Physiological Anthropology, 33(1), 2. https://doi.org/10.1186/1880-6805-33-2

143. Sharma, M., Wasan, A., & Sharma, R. K. (2021). Recent developments in probiotics: An emphasis on Bifidobacterium. Food Bioscience, 41, 100993. https://doi.org/10.1016/j.fbio.2021.100993

144. Shin, J. M., Luo, T., Lee, K. H., Guerreiro, D., Botero, T. M., McDonald, N. J., & Rickard, A. H. (2018). Deciphering Endodontic Microbial Communities by Next-generation Sequencing. Journal of Endodontics, 44(7), 1080–1087. https://doi.org/10.1016/j.joen.2018.04.003

145. Singhania, R. R., Patel, A. K., Soccol, C. R., & Pandey, A. (2009). Recent advances in solid-state fermentation. Biochemical Engineering Journal, 44(1), 13–18. https://doi.org/10.1016/j.bej.2008.10.019

146. Siro, I. (2012). Active and Intelligent Packaging of Food. In Progress in Food Preservation (pp. 23–48). John Wiley & Sons, Ltd. https://doi.org/10.1002/9781119962045.ch2
147. Slatter, W. L., & Halvorson, H. O. (1947). The Heat Resistance of Lactobacilli found in American Cheddar Cheese1. Journal of Dairy Science, 30(4), 231–243. https://doi.org/10.3168/jds.S0022-0302(47)92343-6
148. Sleator, R. D., & Hill, C. (2008). 'Bioengineered Bugs' – A patho-biotechnology approach to probiotic research and applications. Medical Hypotheses, 70(1), 167–169. https://doi.org/10.1016/j.mehy.2007.03.008
149. Soltani, S., Couture, F., Boutin, Y., Ben Said, L., Cashman-Kadri, S., Subirade, M., … Fliss, I. (2021). In vitro investigation of gastrointestinal stability and toxicity of 3-hyrdoxypropionaldehyde (reuterin) produced by Lactobacillus reuteri. Toxicology Reports, 8, 740–746. https://doi.org/10.1016/j.toxrep.2021.03.025
150. Stepek, I. A., & Bode, J. W. (2018). Synthetic fermentation of bioactive molecules. Current Opinion in Chemical Biology, 46, 18–24. https://doi.org/10.1016/j.cbpa.2018.03.014
151. Stratton, J., Chiruvolu, V., & Meagher, M. (1998). High Cell-Density Fermentation. In D. R. Higgins & J. M. Cregg (Eds.), Pichia Protocols (pp. 107–120). Totowa, NJ: Humana Press. https://doi.org/10.1385/0-89603-421-6:107
152. Sudhakar, M. P., Merlyn, R., Arunkumar, K., & Perumal, K. (2016). Characterization, pretreatment and saccharification of spent seaweed biomass for bioethanol production using baker's yeast. Biomass and Bioenergy, 90, 148–154. https://doi.org/10.1016/j.biombioe.2016.03.031
153. Sunesen, L. O., & Stahnke, L. H. (2003). Mould starter cultures for dry sausages—Selection, application and effects. Meat Science, 65(3), 935–948. https://doi.org/10.1016/S0309-1740(02)00281-4
154. Swiegers, J. H., Willmott, R., Hill-Ling, A., Capone, D. L., Pardon, K. H., Elsey, G. M., … Pretorius, I. S. (2006). Modulation of volatile thiol and ester aromas by modified wine yeast. In W. L. P. Bredie & M. A. Petersen (Eds.), Developments in Food Science (pp. 113–116). Elsevier. https://doi.org/10.1016/S0167-4501(06)80027-0
155. Tamang, Jyoti P., Holzapfel, W. H., Shin, D. H., & Felis, G. E. (2017). Editorial: Microbiology of Ethnic Fermented Foods and Alcoholic

Beverages of the World. Frontiers in Microbiology, 8. Retrieved from https://www.frontiersin.org/articles/10.3389/fmicb.2017.01377

156. Tamang, Jyoti P., Tamang, B., Schillinger, U., Franz, C. M. A. P., Gores, M., & Holzapfel, W. H. (2005). Identification of predominant lactic acid bacteria isolated from traditionally fermented vegetable products of the Eastern Himalayas. International Journal of Food Microbiology, 105(3), 347–356. https://doi.org/10.1016/j.ijfoodmicro.2005.04.024
157. Tamang, Jyoti Prakash. (2010). Diversity of Fermented Foods. In Fermented Foods and Beverages of the World. CRC Press.
158. Tamang, Jyoti Prakash, Cotter, P. D., Endo, A., Han, N. S., Kort, R., Liu, S. Q., … Hutkins, R. (2020). Fermented foods in a global age: East meets West. Comprehensive Reviews in Food Science and Food Safety, 19(1), 184–217. https://doi.org/10.1111/1541-4337.12520
159. Tamura, M., Nakatsuka, T., Tada, M., Kawasaki, Y., Kikuchi, E., & Okai, H. (1989). The Relationship between Taste and Primary Structure of “Delicious Peptide” (Lys-Gly-Asp-Glu-Glu-Ser-Leu-Ala) from Beef Soup. Agricultural and Biological Chemistry, 53(2), 319–325. https://doi.org/10.1080/00021369.1989.10869317
160. Taniguchi, M., Aida, R., Saito, K., Ochiai, A., Takesono, S., Saitoh, E., & Tanaka, T. (2019). Identification and characterization of multifunctional cationic peptides from traditional Japanese fermented soybean Natto extracts. Journal of Bioscience and Bioengineering, 127(4), 472–478. https://doi.org/10.1016/j.jbiosc.2018.09.016
161. Tanruean, K., Penkhrue, W., Kumla, J., Suwannarach, N., & Lumyong, S. (2021). Valorization of Lignocellulosic Wastes to Produce Phytase and Cellulolytic Enzymes from a Thermophilic Fungus, Thermoascus aurantiacus SL16W, under Semi-Solid State Fermentation. Journal of Fungi, 7(4), 286. https://doi.org/10.3390/jof7040286
162. Tappy, L., & Rosset, R. (2017). Fructose Metabolism from a Functional Perspective: Implications for Athletes. Sports Medicine, 47(1), 23–32. https://doi.org/10.1007/s40279-017-0692-4
163. Thomsen, M. H. (2005). Complex media from processing of agricultural crops for microbial fermentation. Applied Microbiology and Biotechnology, 68(5), 598–606. https://doi.org/10.1007/s00253-005-0056-0
164. Tieking, M., & Gänzle, M. G. (2005). Exopolysaccharides from cereal-associated lactobacilli. Trends in Food Science & Technology, 16(1), 79–84. https://doi.org/10.1016/j.tifs.2004.02.015

165. Tolve, R., Tchuenbou-Magaia, F., Di Cairano, M., Caruso, M. C., Scarpa, T., & Galgano, F. (2021). Encapsulation of bioactive compounds for the formulation of functional animal feeds: The biofortification of derivate foods. Animal Feed Science and Technology, 279, 115036. https://doi.org/10.1016/j.anifeedsci.2021.115036
166. Tomita, S., Nakamura, T., & Okada, S. (2018). NMR- and GC/MS-based metabolomic characterization of sunki, an unsalted fermented pickle of turnip leaves. Food Chemistry, 258, 25–34. https://doi.org/10.1016/j.foodchem.2018.03.038
167. Tou, E. H., Mouquet-Rivier, C., Rochette, I., Traoré, A. S., Trèche, S., & Guyot, J. P. (2007). Effect of different process combinations on the fermentation kinetics, microflora and energy density of ben-saalga, a fermented gruel from Burkina Faso. Food Chemistry, 100(3), 935–943. https://doi.org/10.1016/j.foodchem.2005.11.007
168. Tran, Q. H., & Unden, G. (1998). Changes in the proton potential and the cellular energetics of Escherichia coli during growth by aerobic and anaerobic respiration or by fermentation. European Journal of Biochemistry, 251(1–2), 538–543. https://doi.org/10.1046/j.1432-1327.1998.2510538.x
169. Tuohy, K. M., Hinton, D. J. S., Davies, S. J., Crabbe, M. J. C., Gibson, G. R., & Ames, J. M. (2006). Metabolism of Maillard reaction products by the human gut microbiota – implications for health. Molecular Nutrition & Food Research, 50(9), 847–857. https://doi.org/10.1002/mnfr.200500126
170. Tyc, Olaf, Song, C., Dickschat, J. S., Vos, M., & Garbeva, P. (2017). The Ecological Role of Volatile and Soluble Secondary Metabolites Produced by Soil Bacteria. Trends in Microbiology, 25(4), 280–292. https://doi.org/10.1016/j.tim.2016.12.002
171. Van Der Borght, A., Goesaert, H., Veraverbeke, W. S., & Delcour, J. A. (2005). Fractionation of wheat and wheat flour into starch and gluten: Overview of the main processes and the factors involved. Journal of Cereal Science, 41(3), 221–237. https://doi.org/10.1016/j.jcs.2004.09.008
172. Wang, F.-S., Su, T.-L., & Jang, H.-J. (2001). Hybrid Differential Evolution for Problems of Kinetic Parameter Estimation and Dynamic Optimization of an Ethanol Fermentation Process. Industrial & Engineering Chemistry Research, 40(13), 2876–2885. https://doi.org/10.1021/ie000544+

173. Wang, Y., Tian, T., Zhao, J., Wang, J., Yan, T., Xu, L., … Zhou, S. (2012). Homofermentative production of d-lactic acid from sucrose by a metabolically engineered Escherichia coli. Biotechnology Letters, 34(11), 2069–2075. https://doi.org/10.1007/s10529-012-1003-7
174. Webb, A. D., & Staff, U. by. (2007). Vinegar. In Kirk-Othmer Encyclopedia of Chemical Technology. John Wiley & Sons, Ltd. https://doi.org/10.1002/0471238961.2209140523050202.a01.pub2
175. Witzig, R., Manz, W., Rosenberger, S., Kruger, U., Kraume, M., & Szewzyk, U. (2002). Microbiological aspects of a bioreactor with submerged membranes for aerobic treatment of municipal wastewater. Water Research, 36(2), 394–402. https://doi.org/10.1016/S0043-1354(01)00221-4
176. Wu, C.-W., Chen, R.-H., Pu, J.-Y., & Lin, T.-H. (2004). The influence of air–fuel ratio on engine performance and pollutant emission of an SI engine using ethanol–gasoline-blended fuels. Atmospheric Environment, 38(40), 7093–7100. https://doi.org/10.1016/j.atmosenv.2004.01.058
177. Wu, R., Yu, M., Liu, X., Meng, L., Wang, Q., Xue, Y., & Yue, X. (2015). Changes in flavour and microbial diversity during natural fermentation of suan-cai, a traditional food made in Northeast China. International Journal of Food Microbiology, 211, 23–31. https://doi.org/10.1016/j.ijfoodmicro.2015.06.028
178. Xie, F., Quan, S., Liu, D., Ma, H., Li, F., Zhou, F., & Chen, G. (2014). Purification and characterization of a novel α-amylase from a newly isolated Bacillus methylotrophicus strain P11-2. Process Biochemistry, 49(1), 47–53. https://doi.org/10.1016/j.procbio.2013.09.025
179. Yamagishi, H., & Ogata, T. (1999). Chromosomal Structures of Bottom Fermenting Yeasts. Systematic and Applied Microbiology, 22(3), 341–353. https://doi.org/10.1016/S0723-2020(99)80041-1
180. Yang, B., Chen, J., Yu, Q., Li, H., Lin, M., Mustapha, A., & Wang, Y. (2011). Oral bacterial deactivation using a low-temperature atmospheric argon plasma brush. Journal of Dentistry, 39(1), 48–56. https://doi.org/10.1016/j.jdent.2010.10.002
181. Ye, R. W., Wang, T., Bedzyk, L., & Croker, K. M. (2001). Applications of DNA microarrays in microbial systems. Journal of Microbiological Methods, 47(3), 257–272. https://doi.org/10.1016/S0167-7012(01)00308-6

182. Yeh, C.-M., Kao, B.-Y., & Peng, H.-J. (2009). Production of a Recombinant Type 1 Antifreeze Protein Analogue by L. lactis and Its Applications on Frozen Meat and Frozen Dough. Journal of Agricultural and Food Chemistry, 57(14), 6216–6223. https://doi.org/10.1021/jf900924f

183. Yin, L., Wang, C., Zhu, X., Ning, C., Gao, L., Zhang, J., & Huang, R. (2020). A multi-step screening approach of suitable non-Saccharomyces yeast for the fermentation of hawthorn wine. LWT, 127, 109432. https://doi.org/10.1016/j.lwt.2020.109432

184. Yuan, Y., Feng, F., Chen, L., Yao, Q., & Chen, K. (2013). Directional isolation of ethanol-tolerant acetic acid bacteria from industrial fermented vinegar. European Food Research and Technology, 236(3), 573–578. https://doi.org/10.1007/s00217-012-1885-6

185. Zamora, F. (2009). Biochemistry of Alcoholic Fermentation. In M. V. Moreno-Arribas & M. C. Polo (Eds.), Wine Chemistry and Biochemistry (pp. 3–26). New York, NY: Springer. https://doi.org/10.1007/978-0-387-74118-5_1

186. Zhao, Y., Li, A., Gao, R., Tao, P., & Shen, J. (2014). Measurement of temperature, relative humidity and concentrations of CO, CO2 and TVOC during cooking typical Chinese dishes. Energy and Buildings, 69, 544–561. https://doi.org/10.1016/j.enbuild.2013.11.037

187. Zhu, Y., & Tramper, J. (2013). Koji – where East meets West in fermentation. Biotechnology Advances, 31(8), 1448–1457. https://doi.org/10.1016/j.biotechadv.2013.07.001

188. Zuniga, M., Pardo, I., & Ferrer, S. (1993). An improved medium for distinguishing between homofermentative and heterofermentative lactic acid bacteria. International Journal of Food Microbiology, 18(1), 37–42. https://doi.org/10.1016/0168-1605(93)90005-2

INDEX

D

E

S

T

U

V

W

X

Y

Z

Printed in the United States
by Baker & Taylor Publisher Services